Atlas of Design

Vol. 3 »

editors
Samuel V. Matthews
Martin E. Elmer

assistant editor
Virginia W. Mason

published by
nacis
the North American
Cartographic Information Society

Atlas of Design, Volume Three

Third Printing, July 2023

Editors: Samuel V. Matthews and Martin E. Elmer

Assistant Editor: Virginia W. Mason

Published by the North American Cartographic Information Society, *nacis.org*

The type in this book has been set in Avenir, by Adrian Frutiger.

Photos of the *Atlas of Design* are courtesy of Caroline Rose.

Printed in the United States of America by John Roberts, Inc., Minneapolis, MN.

ISBN 979-8-9866856-3-2

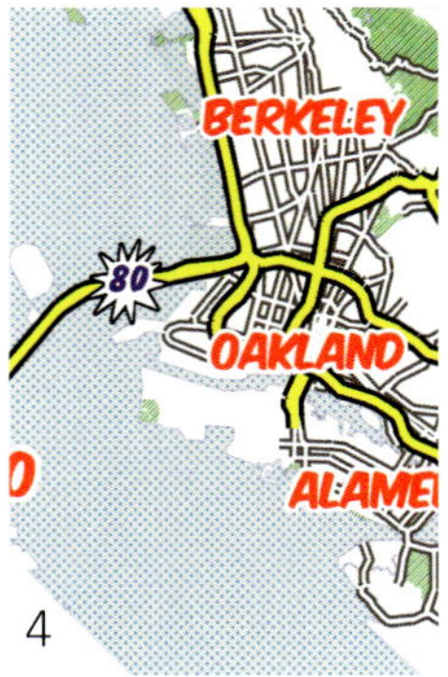

4

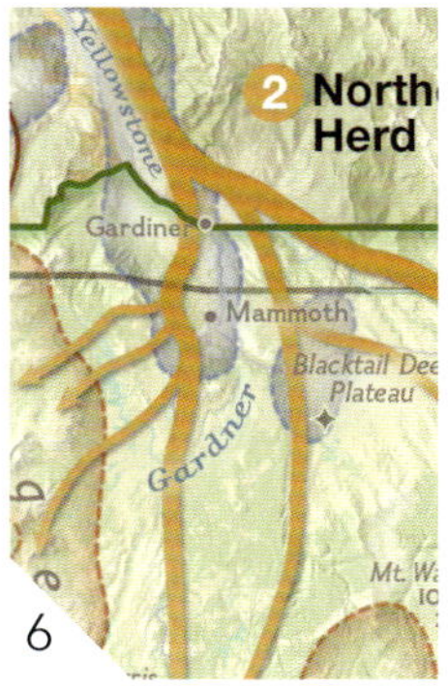

6

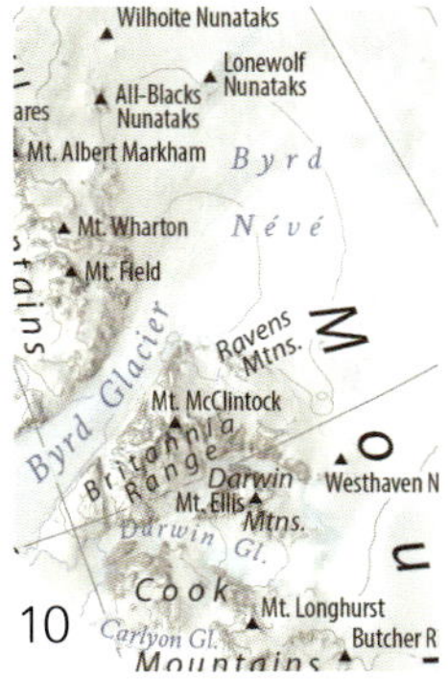

10

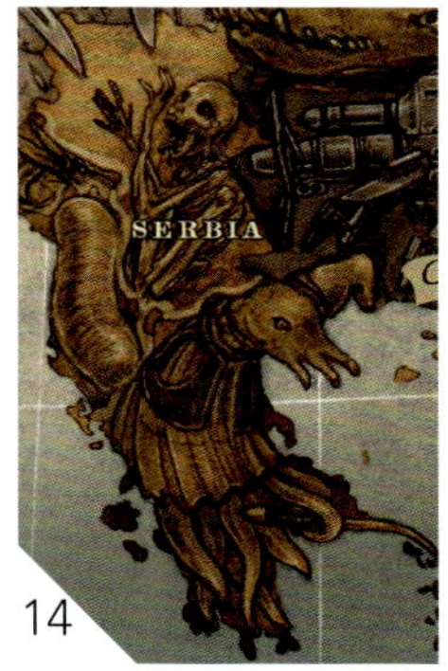

14

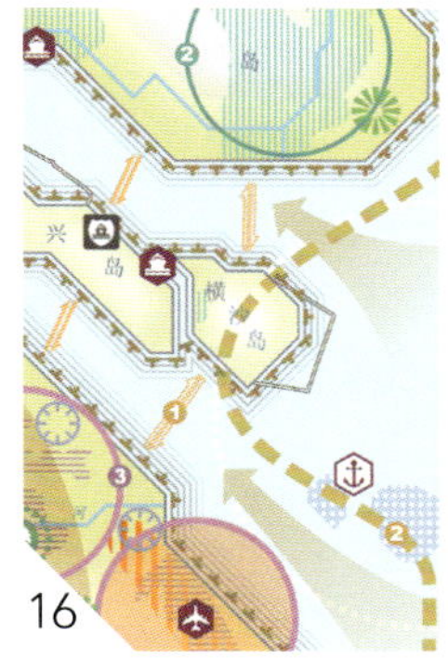

16

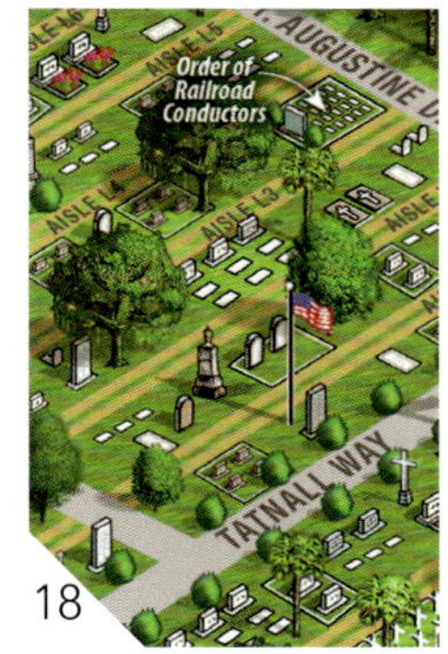

18

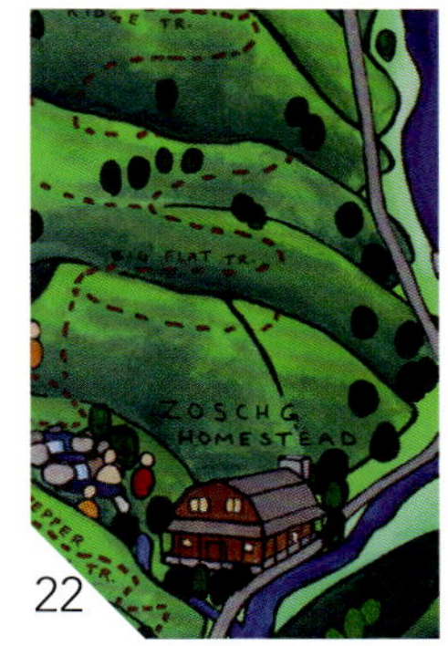

22

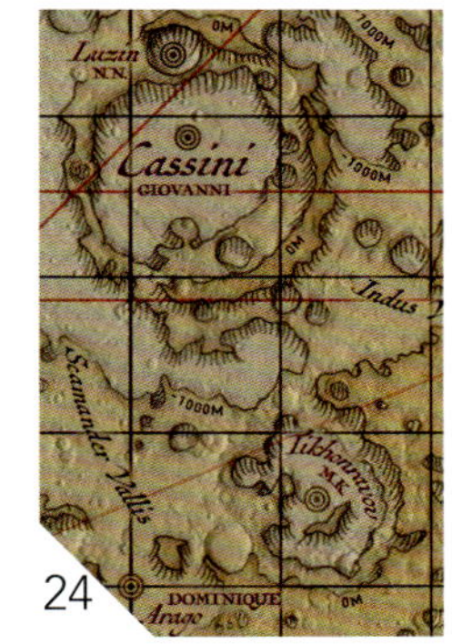

24

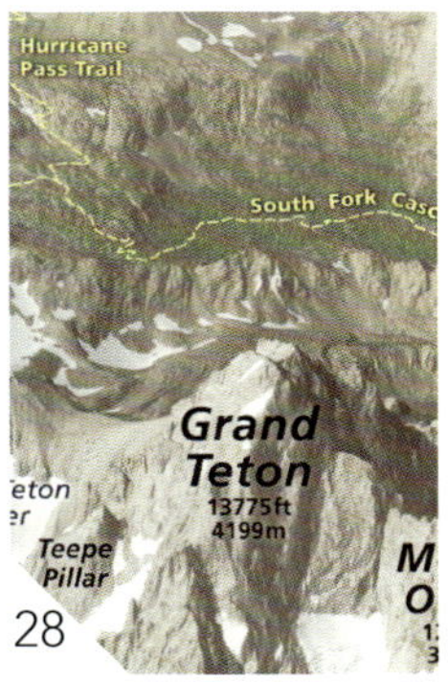

28

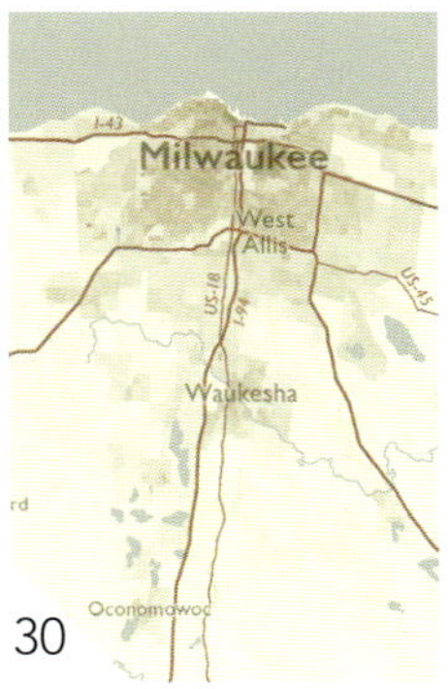

30

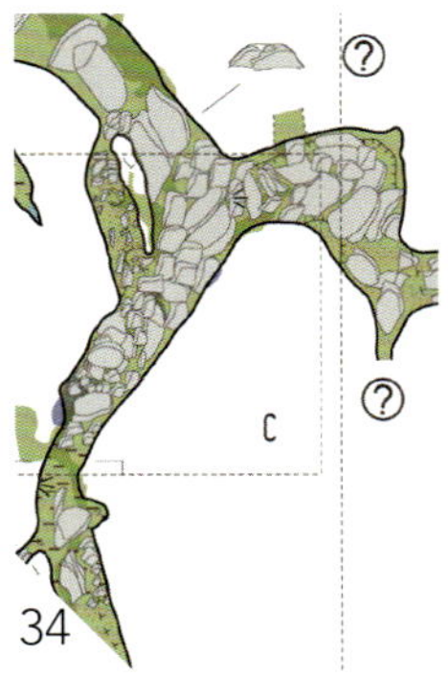

34

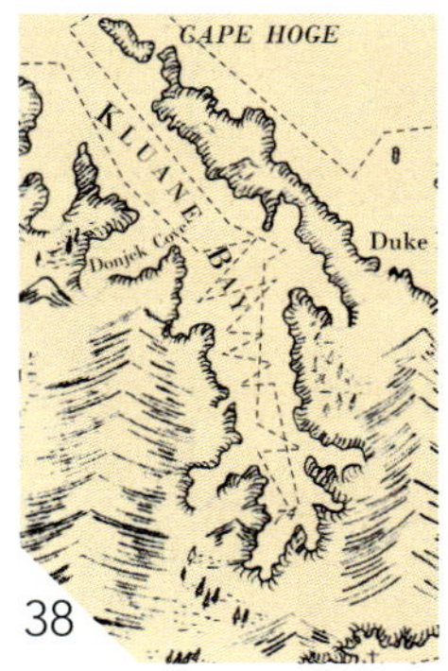

38

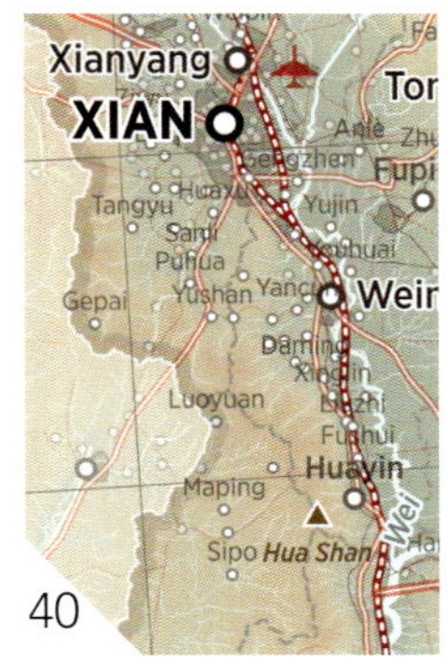

40

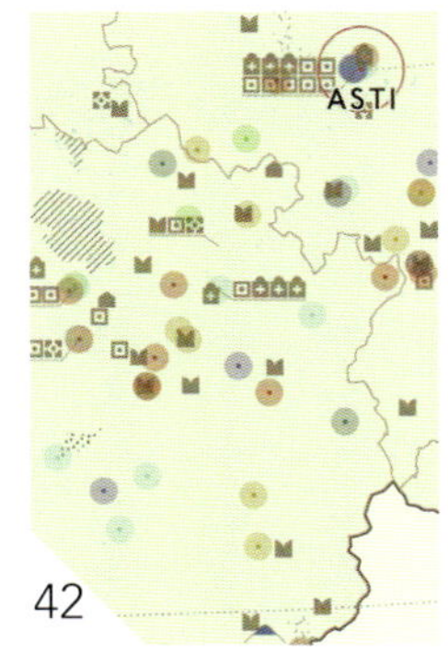

42

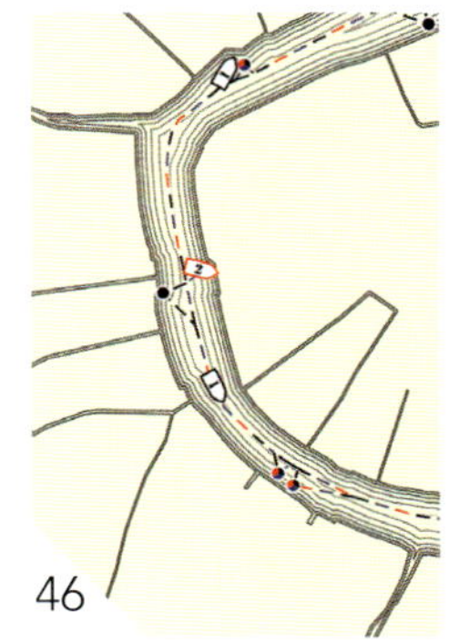

46

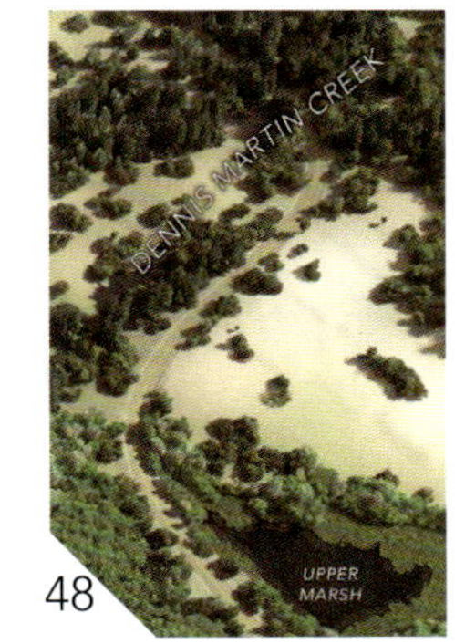

48

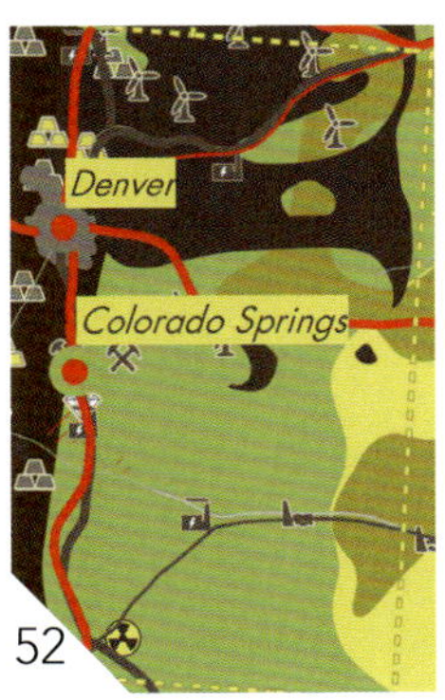

52

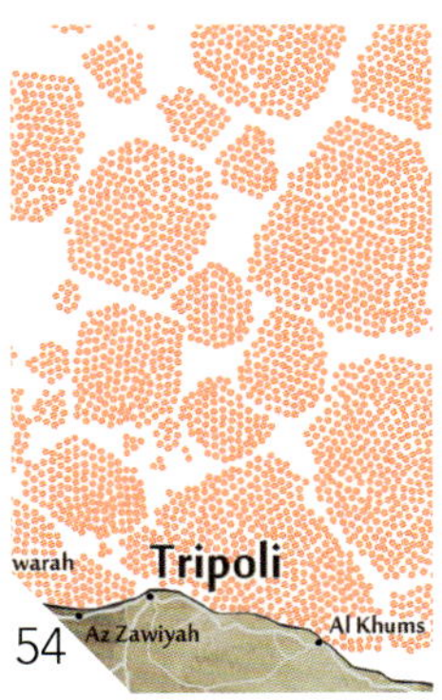

54

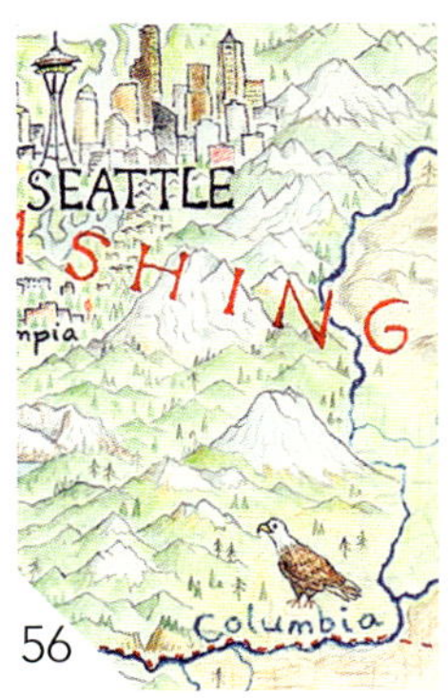

56

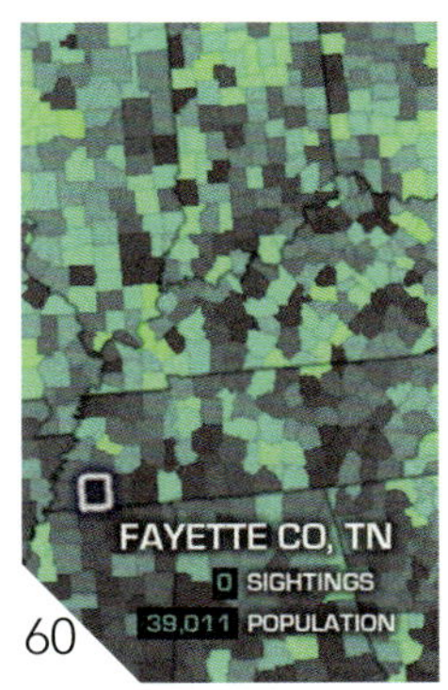

60

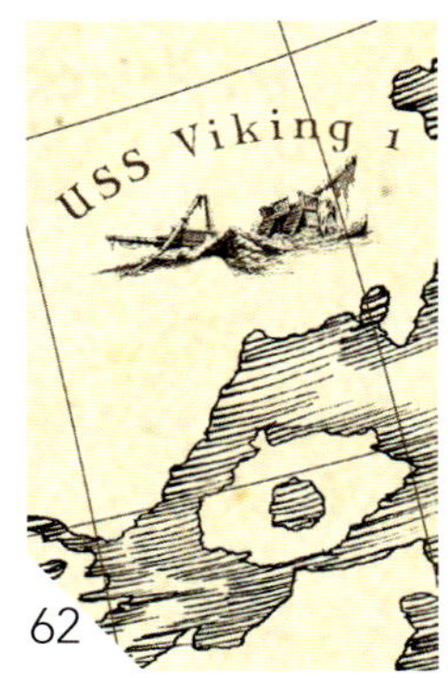

62

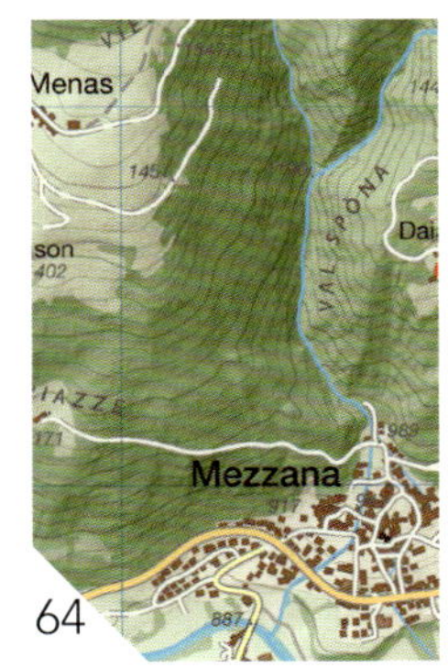

64

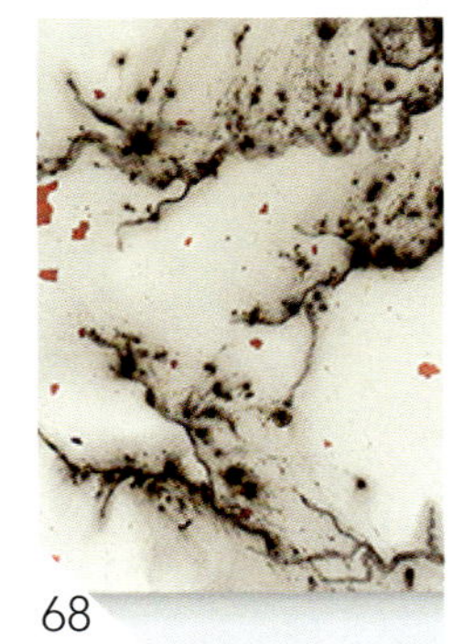

68

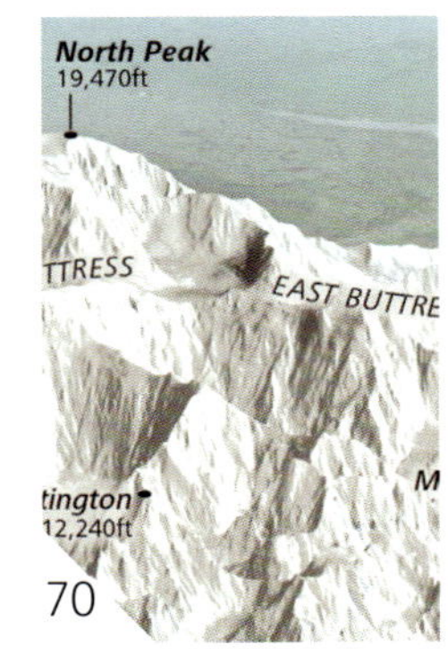

70

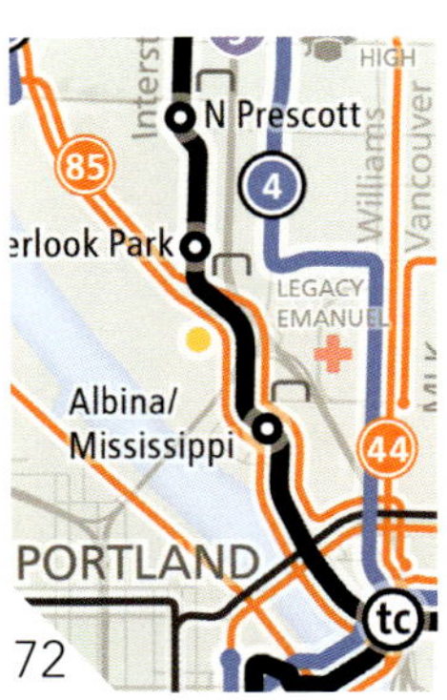

72

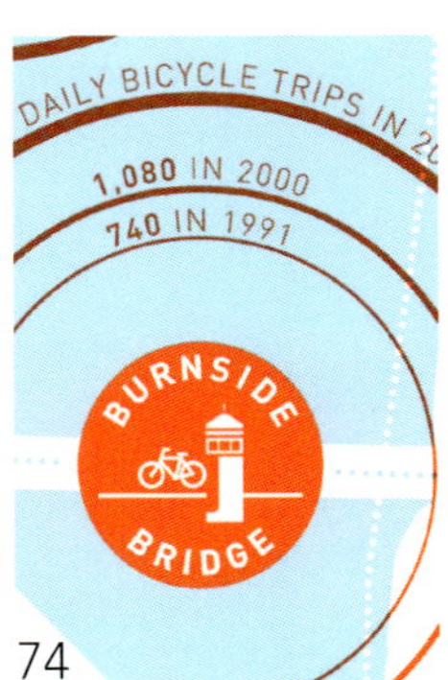

74

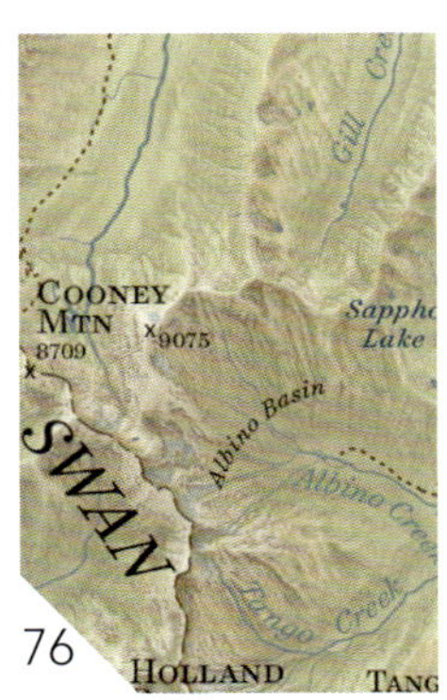

76

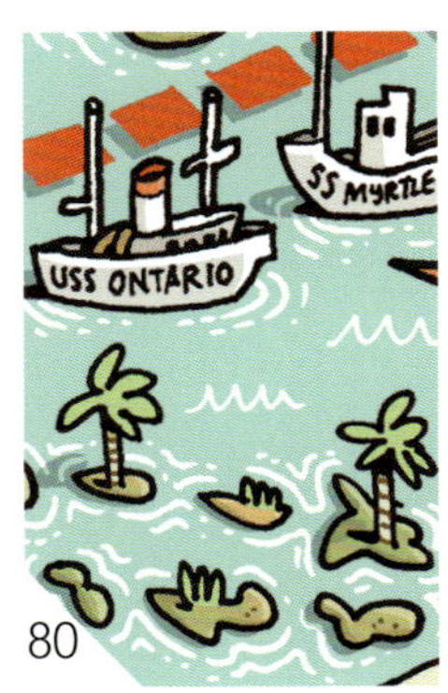

80

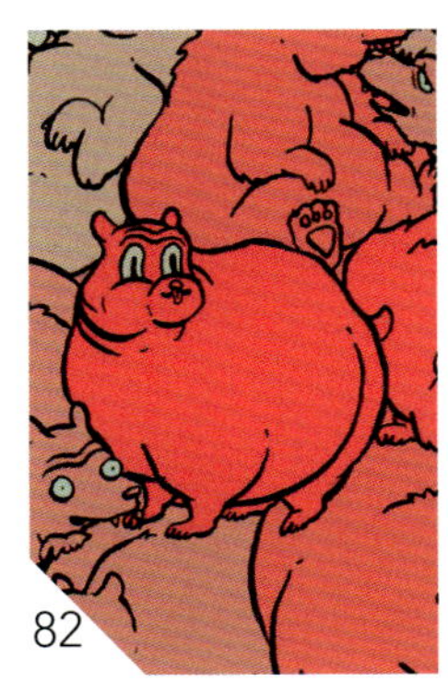

82

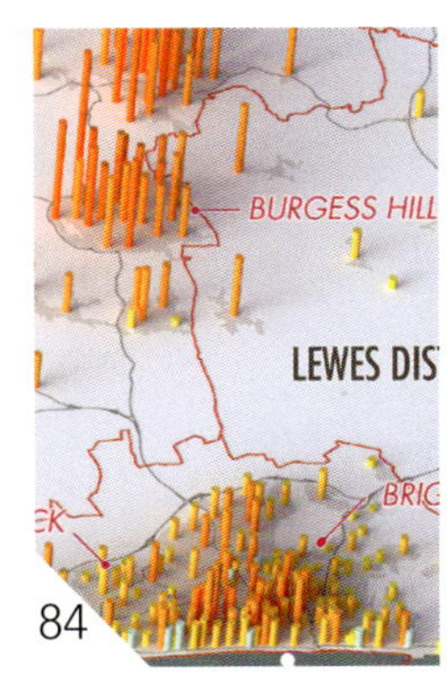

84

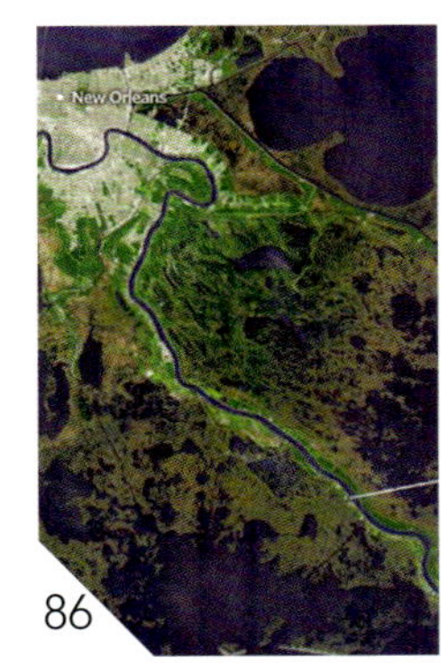

86

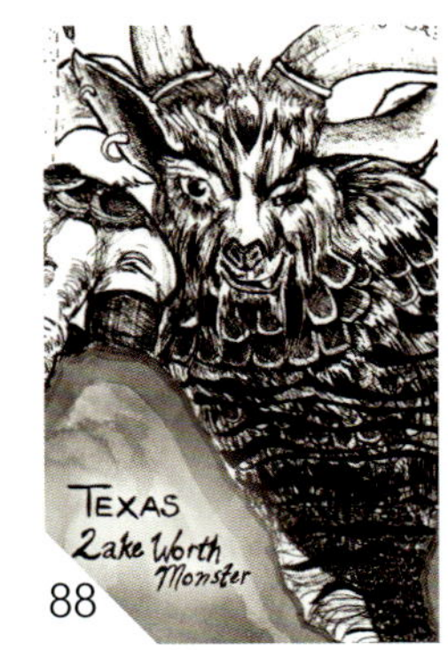

88

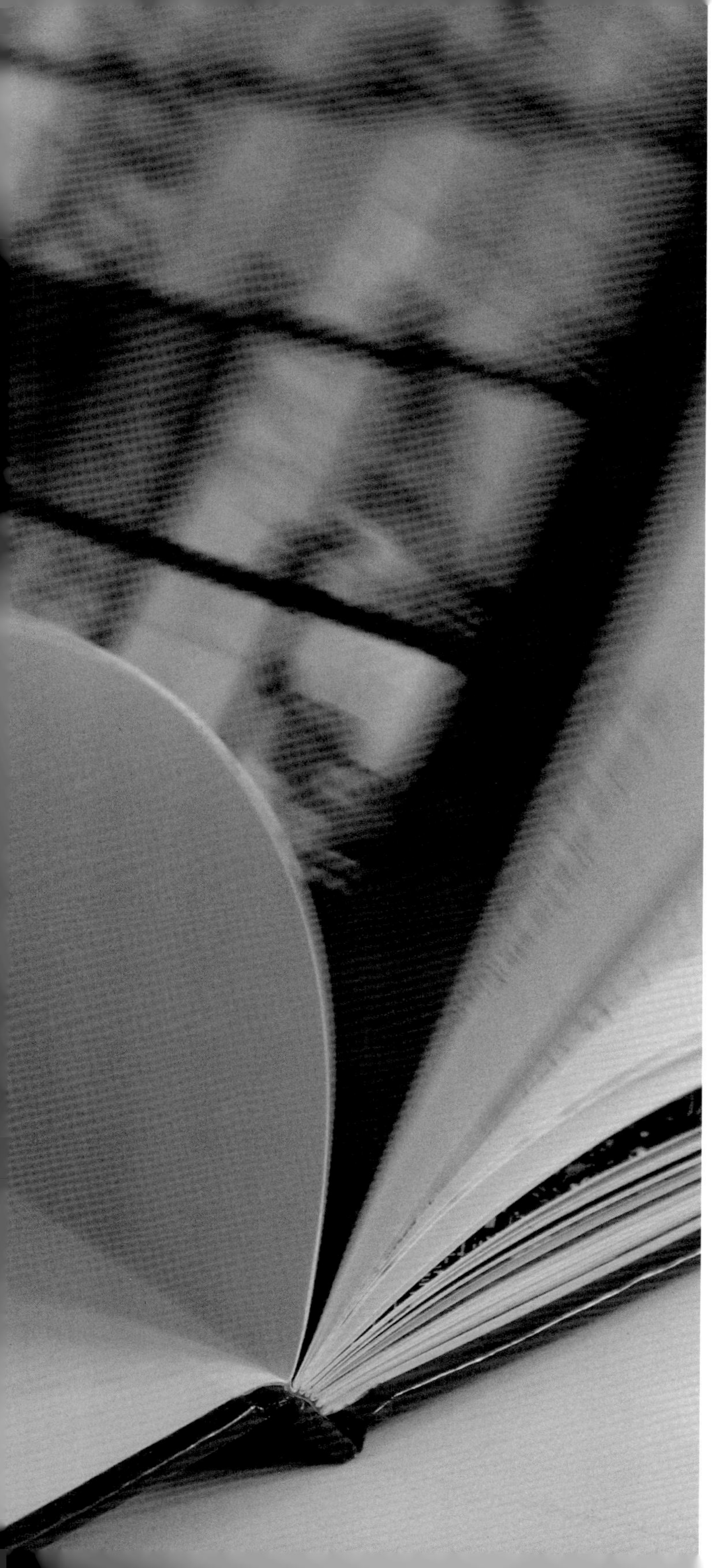

nacis

The *Atlas of Design* is a publication of the North American Cartographic Information Society (NACIS), a community of people with a professional interest in maps and geographic data. NACIS brings together professionals, students, and interested observers in academic, business and government sectors, including cartographers, librarians, and software creators, to advance the cartographic arts through its Annual Meeting and other initiatives.

The Atlas of Design is one such initiative, a gallery of the best of contemporary cartography. Every two years, a volunteer team of NACIS editors choose from hundreds of entries from every corner of the cartographic world. More than a picture book, the atlas is designed to spread a variety of top mapmakers' ideas about what it means to make a good map, and inspire others to follow their lead. We thank the contributors for their part in this process, and congratulate them on their selection.

Please enjoy the end result. To learn more about other ways NACIS is advancing cartography, or to join us, visit nacis.org.

Acknowledgements

This is the first volume of the *Atlas* prepared without the original editors: Daniel Huffman and Tim Wallace. Inheriting a project as well-crafted as the *Atlas* has been a joy, and we are thrilled to introduce the third volume.

The nearly 250 submissions from more than 15 countries provided the *Atlas* with a wealth of inspiration. The authors submitting their work have put in countless hours researching, designing, and perfecting their maps. We thank every participant, featured or not, for their generosity in providing us with wealth of content necessary for this publication.

Reviewing so many maps is no simple task. We are incredibly grateful for the panel of judges willing to donate their time helping us narrow the submissions down to the final works. Many thanks to AJ Ashton, Alethea Steingisser, Andy Woodruff, Daniel Huffman, Hans van der Maarel, Ian Muehlenhaus, Margaret Pearce, Nat Case, Rob Roth, Tanya Buckingham, Tim Wallace, and Tom Patterson.

None of us, as editors, are qualified, nor have the ability to print so many books at such quality. The American Printing Company has been our partner from the beginning and we are most thoroughly grateful for their expertise and continued interest in the *Atlas*.

Finally, we thank you, the reader. The 32 maps in the following pages were created in different places by very different authors. They are meant to teach you about the world and to transport you into the unknown. Many may seem familiar, and yet completely new. These maps—these pages—are meant for you. Thanks for supporting the *Atlas of Design*.

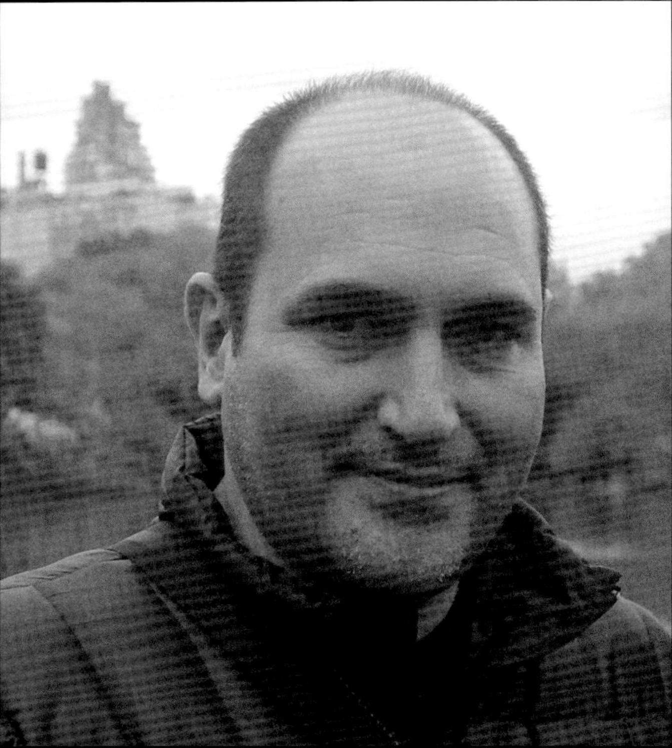

Introduction by Juan Velasco »

Founder & Creative Director of 5W Graphics and former Art Director of *National Geographic* Magazine

When I became art director of *National Geographic* magazine in 2008 I found myself, all of a sudden, in charge of the best cartography department in the world. Me? I couldn't believe it. I soon discovered that cartographers are a peculiar bunch, and for a while they put the fear of God in me.

From my first day at *National Geographic*, I was in awe of the cartographers' incredible ability to visualize and explain the world. And I admit that I was slightly intimidated by my new team—a group of men and women devoted to beauty and precision, artists whose tools were shapefiles, DEMs, and bathymetry. Their choropleth-orthographic-Winkel-Tripel-Projection- obliques were incomprehensible at first. These were Real Cartographers of the kind I had never encountered, despite years of "mapping" work. I knew they would grant me no mercy when they saw the nefarious mapping standards to which I was accustomed.

I was and am a journalist specializing in information graphics. I've enjoyed a long career in newspapers, where the most typical mapping technique is tracing on top of a Google Maps screengrab or over the tourist map of Dublin that Carlos, the metro section reporter, brought home from vacation. And yes, this time-honored technique is still used, with rare exceptions. The creative editing we have to employ when inescapable, in-your-face deadlines rule our lives means that important physical features are often eliminated from our maps without a second thought.

But in spite of our differences, as I worked with expert cartographers on project after project, I discovered how much we had in common, especially our shared passion for visual storytelling.

In my youth, I met cartographers through literature—and they were peculiar and awe-inspiring too. In a one-paragraph short story, "On Exactitude in Science," Jorge Luis Borges describes, in the guise of a fictional 17th-century chronicle, a place where the art and science of cartography is so precise and obsessive that only a map on the same scale as an empire will suffice:

. . . *In that Empire, the Art of Cartography attained such Perfection that the map of a single Province occupied the entirety of a City, and the map of the Empire, the entirety of a Province. In time, those Unconscionable Maps no longer satisfied, and the Cartographers Guilds struck a Map of the Empire*

In all artistic endeavors, imposing limits on ourselves—and pushing against those limits as hard as possible to seek beauty and truth—is among the most satisfying things we can do.

whose size was that of the Empire, and which coincided point for point with it. The following Generations, who were not so fond of the Study of Cartography as their Forebears had been, saw that that vast Map was Useless, and not without some Pitilessness was it, that they delivered it up to the Inclemencies of Sun and Winters. In the Deserts of the West, still today, there are Tattered Ruins of that Map, inhabited by Animals and Beggars; in all the Land there is no other Relic of the Disciplines of Geography.

—From Suárez Miranda, Travels of Prudent Men, Book IV, Ch. XLV, Lérida, 1658

Borges, who published his story in 1946, was not the first to imagine the concept of a life-size map. In *Sylvie and Bruno Concluded*, written in 1893, Lewis Carroll described such a map too, one that would include every single feature in existence "on the scale of a mile to the mile," so large it would "cover the whole country and shut out the sunlight." And in our own century, Neil Gaiman paid homage to Borges in "The Mapmaker," a story about a Chinese emperor who dreams of a map-world so grand that it is "perfectly accurate, and perfectly useless."

After we rule out the impossible and the absurd, these writers prompt us to ask a serious question: What is the value of any map, if not perfect accuracy? For me, the answer lies in the storytelling purpose of maps—a purpose fulfilled in every map in this book. The storytelling power of each of these maps goes beyond words and images as a result of the maker's point of view. As Borges shows so succinctly, a map that reproduces the world without an artist's interpretation is worthless.

Among the many types of journalistic information graphics—including charts, diagrams, illustrations, and maps—I have always considered maps the most challenging. Charts reveal trends and stories hidden within large sets of numbers, and they have the inherent advantage of the abstraction and simplicity of those numbers. They stand alone, clean and uncomplicated by background noise. In the case of illustrations and diagrams, the information the makers are trying to convey is in the image itself. A cutaway reveals immediately the interior of a machine in literal, recognizable representation. But maps are different.

The irony of a thematic or data-driven map is that the map itself is potentially an obstacle to delivering information clearly. A map often must strive to recede and serve as a background holding the data layer, which delivers the important message to readers. It's easy to show an enormous amount of information in a map. It's extremely difficult to do so while telling a well-structured, well-edited, insightful story that is absolutely clear—and to make it beautiful. Great cartographers know how to use restraint and balance and always have a highly refined sense of design.

After the subtlety of well-rendered shaded relief, the element of cartography I enjoy the most is typography. But typography is just one item in the cartographer's toolbox. A cartographer must be a good reporter and researcher. He or she understands data and statistics; graphic design and color as information; a variety of storytelling techniques; and concise, efficient writing that supports the visuals and advances the reader's understanding. Time and again at *National Geographic* I saw how the maps team used all these tools and more to illuminate the world for our readers.

The true beauty of mapmaking lies in constant struggle between the absolute necessity of accuracy and the inevitability of abstraction. At the same time, mapmakers feel an equally intense struggle between the need to communicate a story efficiently and the impulse to use their boundless creativity to deliver visual beauty. These daily struggles keep our work engaging and alive. In all artistic endeavors, imposing limits on ourselves—and pushing against those limits as hard as possible to seek beauty and truth—is among the most satisfying things we can do.

In the end we must return to storytelling. French philosopher Gilles Deleuze said, "Writing has nothing to do with meaning. It has to do with land surveying and cartography, including the mapping of countries yet to come." I see no difference between a good infographics editor, a good cartographer, and a good writer. We all begin with uncharted territories and a blank page; then we explore the landscape and tell our stories.

Today, I still stand in awe of cartographers, although I'm not as frightened as I once was. The printed and digital maps they make, much like the mental maps that guide our lives, are constant reminders of our human need for reference and understanding—all the better if the maps are beautiful as well as informative.

This book is a formidable compilation of beautiful and informative cartography, from classically designed to bold and daring; from hand-drawn gems to multilayered interactive presentations. It is a collection of short stories about the world we live in, each as fascinating and unique as the cartographers who made them.

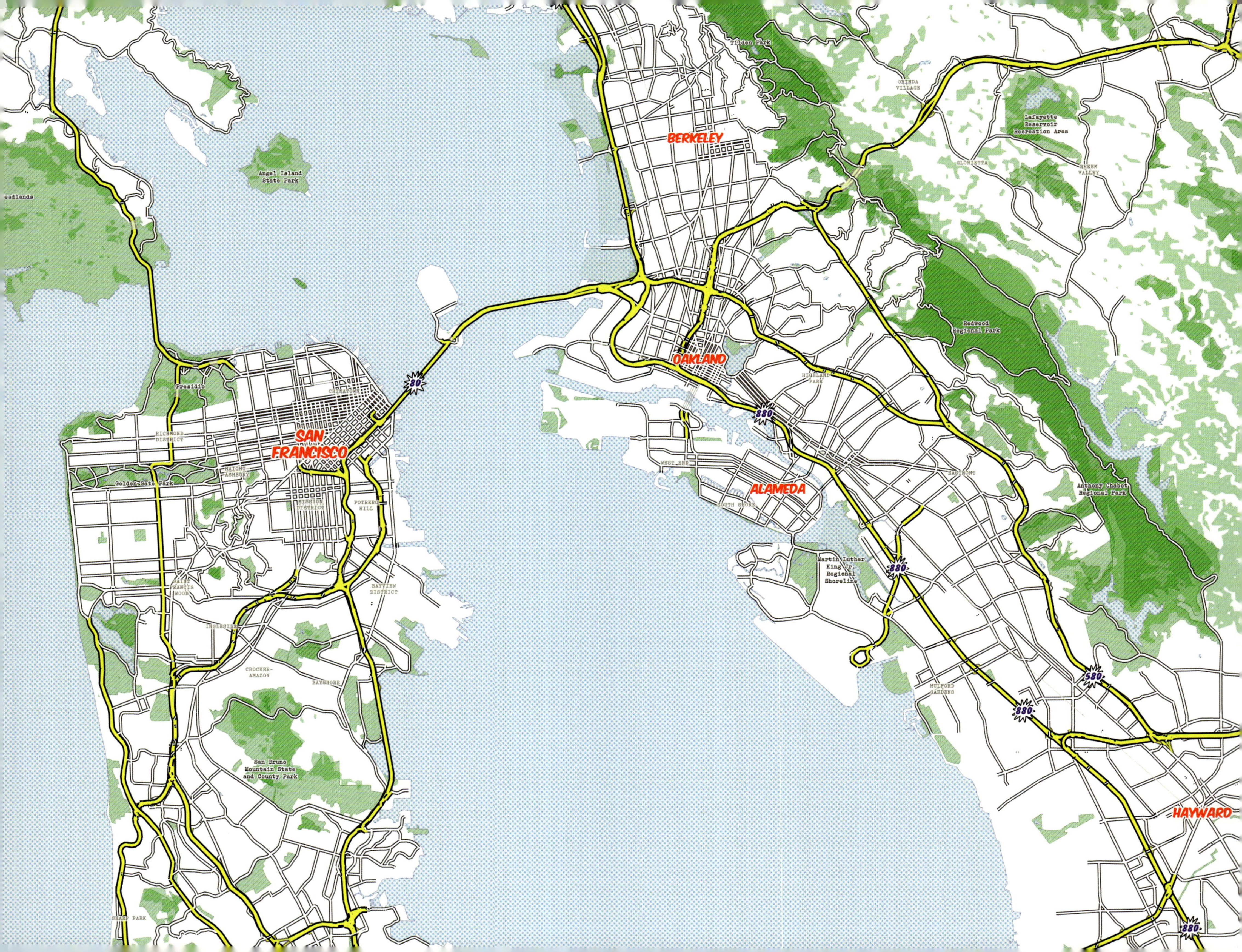
Tilden Park
ORINDA VILLAGE
BERKELEY
Lafayette Reservoir Recreation Area
GLORIETTA
RHEEM VALLEY
Angel Island State Park
Redwood Regional Park
OAKLAND
HIGHLAND PARK
Presidio
CHINATOWN
80
RICHMOND DISTRICT
SAN FRANCISCO
HAIGHT-ASHBURY
Golden Gate Park
MISSION DISTRICT
POTRERO HILL
880
WEST END
ALAMEDA
SOUTH SHORE
EASTMONT
Anthony Chabot Regional Park
Martin Luther King Jr. Regional Shoreline
SAINT FRANCIS WOOD
BAYVIEW DISTRICT
INGLESIDE
CROCKER-AMAZON
BAYSHORE
MULFORD GARDENS
580
San Bruno Mountain State and County Park
HAYWARD

《 Whaam!

Amy Lee Walton | Mapbox

The Pop Art movement resonated with me as a young designer. Pop was a cheeky commentary on American culture and mass production. Was it an artist's take on graphic design or a graphic designer's take on art?

Pop artists created work that mimicked machine-made objects with hand strokes, but today the computer has replaced the hand. We create web and mobile maps with boundless zoom using a mouse and GeoJSON coordinates instead of using brushes and ink presses.

Inspired by Roy Lichtenstein's "Artist's Studio No. 1 (Look Mickey)," my goal was to design a map that captured the whimsy and texture of Pop Art. I planned to capture Lichtenstein's comic book aesthetic by designing with a limited, vibrant color palette; thick black line strokes; Lichtenstein's signature Ben-Day dot texture; starbursts; and a slightly comic display font.

To achieve the Ben-Day dot texture, I put small, closely spaced dark cyan dots over a solid, light cyan background. In Pop Art fashion, I wanted to create a slight optical illusion when zooming in on the map. I incorporated primary and secondary CMYK colors to hint at traditional printing, simultaneously applying red, blue, cyan, green, and yellow variations for a vibrant and playful visual read.

I added a subtle off-white bit of texture to the map for the land masses. When zooming closer, a vibrant green line pattern that's another nod to Lichtenstein's style defines green spaces. For cityscapes, I applied a heavy black stroke to road networks, from highways to main streets, varying the thickness. I added a touch of color to distinguish motorways with a bright yellow fill and thick black stroke, leaving other roads with white fill. Finally, I added a touch of Pop whimsy to the map with a handwritten-style display font in bright red for major city labels and numbers in highway shields in royal blue.

Data © OpenStreetMap osm.org/copyright

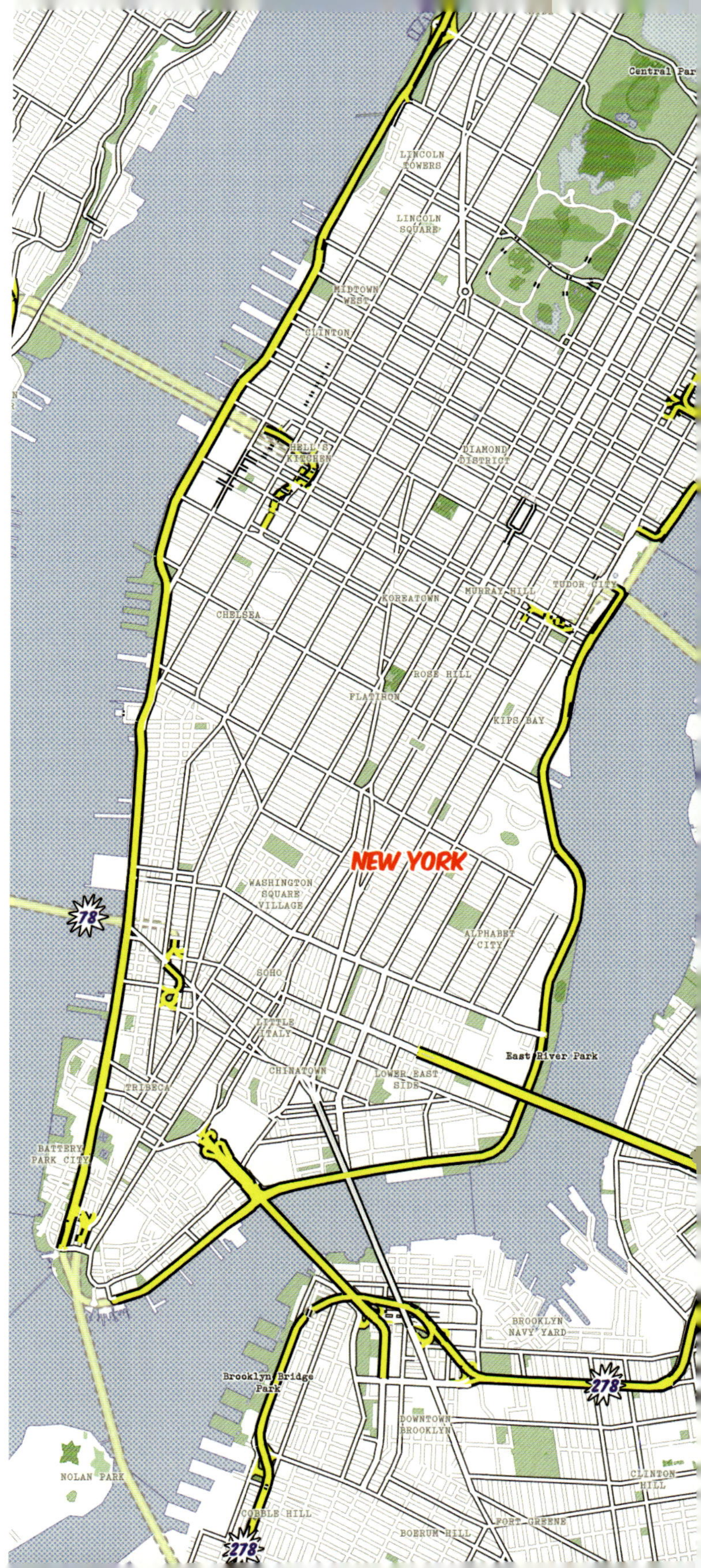

Yellowstone Elk Migration »

National Geographic | Martin Gamache, Arthur Middleton, Lawson Parker, Shelley Sperry, Lauren Tierney

It's no wonder that the Crow and other native tribes knew the Yellowstone River as the Elk River. Elk herds define the Greater Yellowstone ecosystem and support the vast carnivore complex for which the region is known. Our map tells the story of these herds' migrations by weaving together historical references, elk-based place names, archaeological sites, and information about the elk's primary predators: wolves and humans.

The main source of information for this map was ecologist Arthur Middleton, who has compiled a dataset of elk GPS tracks spanning decades. Here, for the first time, Middleton used this tracking data to show herds from around the Greater Yellowstone region moving into and out of Yellowstone National Park on an annual migration driven by the high quality grazing found at higher elevations there during the summer months. We also teamed up with scientists and graduate students from the Wyoming Migration Initiative and fellow cartographers from the University of Oregon InfoGraphics Lab to re-compile the GPS tracks Middleton had assembled. The data was cleaned up, analyzed to determine core winter and summer ranges as well as seasonal start and end dates, and generalized into broad corridors showing the major migration routes.

Yellowstone was a hot topic in 2016 because of the centennial of the National Park Service. Our collaborators and other cartographers were working on smart, informative maps, so when we had all the data in hand, we began to experiment. We agonized over whether to emphasize the many individual elk herds that make up the area's super herd or to emphasize the seasonal movement of elk in and out of Yellowstone. To make matters more challenging, the drama of elk migration plays out over the background of Greater Yellowstone, which is complex in terms of topography and land management, so we used small multiples along the lower edge to reinforce the story of land management in the region.

In the end, we used color to differentiate the known herds, with variations of value and transparency as well as line weight and shape to illustrate the rhythmic, breathing life force of the herds as they move from winter to summer ranges and back.

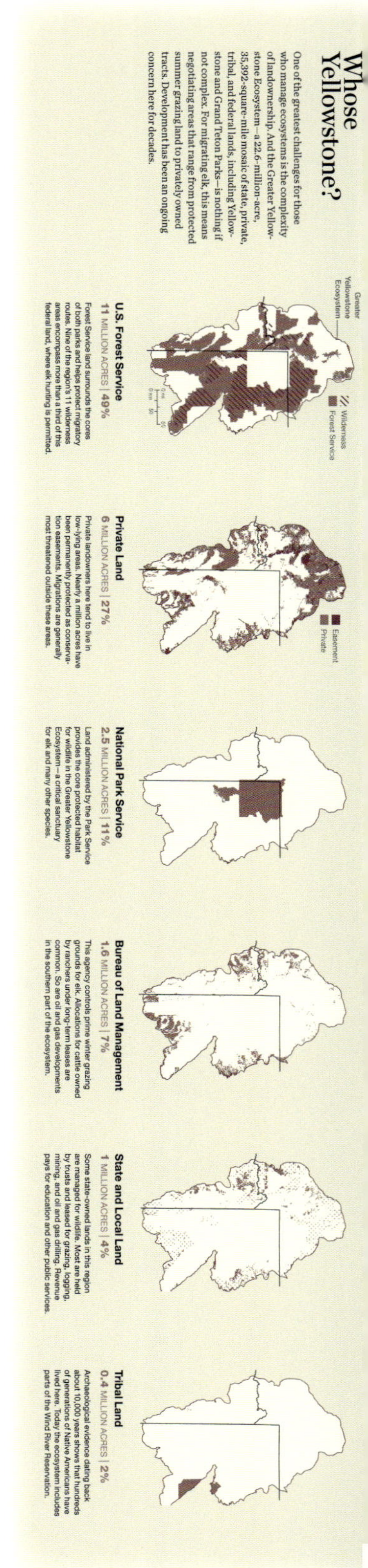

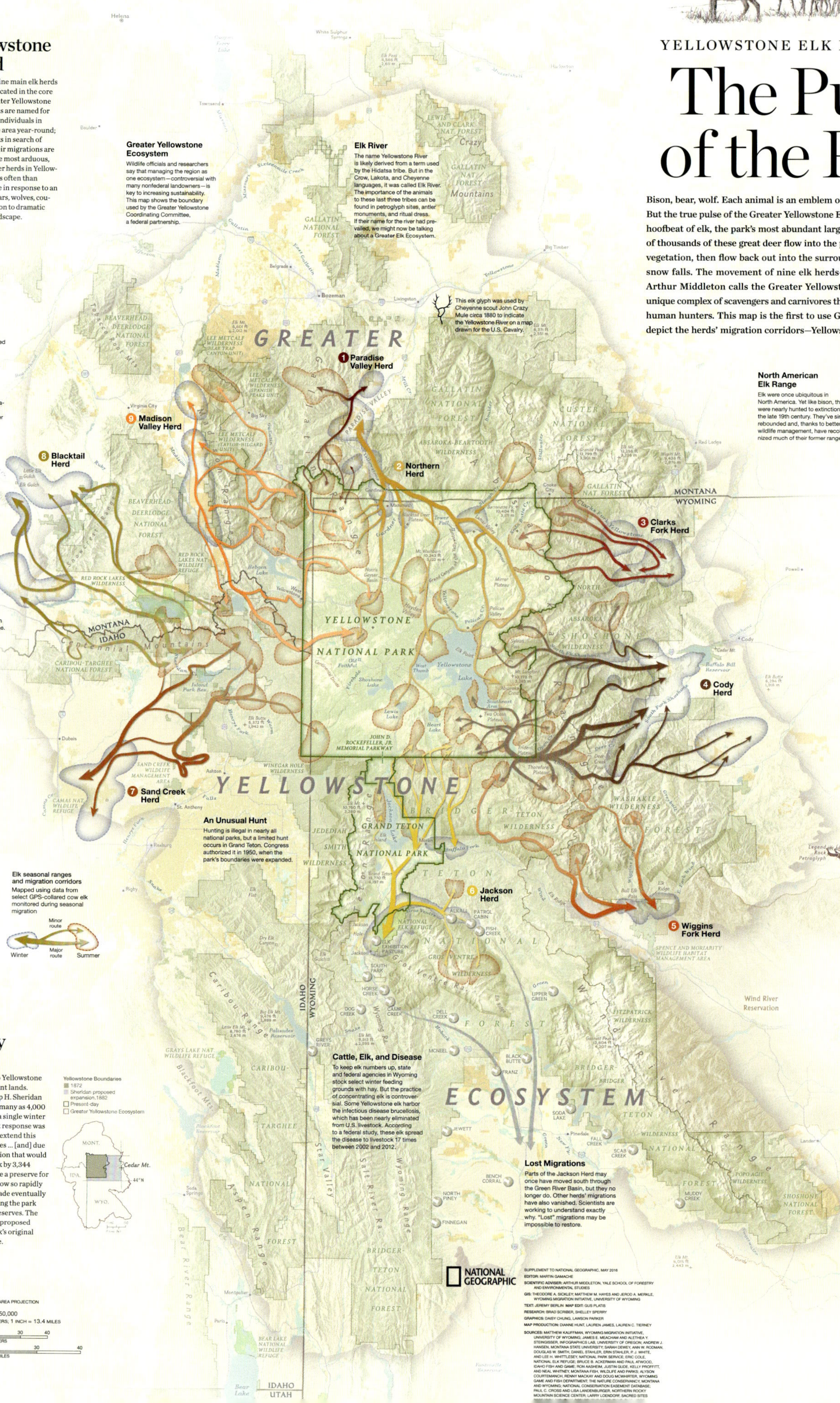

YELLOWSTONE ELK MIGRATIONS

The Pulse of the Park

Bison, bear, wolf. Each animal is an emblem of Yellowstone National Park. But the true pulse of the Greater Yellowstone Ecosystem is measured by the hoofbeat of elk, the park's most abundant large mammal. Each spring tens of thousands of these great deer flow into the park to feed on high-altitude vegetation, then flow back out into the surrounding region when autumn snow falls. The movement of nine elk herds—composing what ecologist Arthur Middleton calls the Greater Yellowstone superherd—supports a unique complex of scavengers and carnivores that, outside the park, includes human hunters. This map is the first to use GPS data from collared elk to depict the herds' migration corridors—Yellowstone's arteries and veins.

The Yellowstone Superherd

This map highlights the nine main elk herds whose summer range is located in the core protected area of the Greater Yellowstone Ecosystem. The nine herds are named for where they winter. Some individuals in each herd stay in the same area year-round; others make seasonal treks in search of better places to graze. Their migrations are among the most stable, the most arduous, and the most studied. Other herds in Yellowstone also migrate, but less often than they once did. That may be in response to an increase in predators—bears, wolves, cougars, and people—in addition to dramatic human changes to the landscape.

1 Paradise Valley
Decades of population growth and development in this agricultural valley have constricted the herd's seasonal migrations.

2 Northern
The only herd that winters in the park is the one most visibly thinned by wolves. Drought, bears, and hunting have also taken a toll.

3 Clarks Fork
More predation and less precipitation have curtailed the migration. Individuals that stay in their winter range find more food and safety.

4 Cody
One of the largest herds follows one of the most rugged routes, from public and private land to a remote part of the ecosystem.

5 Wiggins Fork
A 1991 state land purchase gave this herd even more separation from ranchlands. These elk thrive without supplemental feeding.

6 Jackson
A century of development and expansion in Jackson Hole has reduced this herd, which winters in and around the National Elk Refuge.

7 Sand Creek
This herd has rebounded since the 1950s, when Idaho passed legislation that restored habitat along the elk's migratory route.

8 Blacktail
Living mostly on public land, this herd is highly attractive to hunters. Some of the elk migrate across the Montana-Idaho border.

9 Madison Valley
Huge groups of elk winter on this windy, high-elevation grassland. Their range is largely intact thanks to conservation easements.

Greater Yellowstone Ecosystem
Wildlife officials and researchers say that managing the region as one ecosystem—controversial with many nonfederal landowners—is key to increasing sustainability. This map shows the boundary used by the Greater Yellowstone Coordinating Committee, a federal partnership.

Elk River
The name Yellowstone River is likely derived from a term used by the Hidatsa tribe. But in the Crow, Lakota, and Cheyenne languages, it was called Elk River. The importance of the animals to these last three tribes can be found in petroglyph sites, antler monuments, and ritual dress. If their name for the river had prevailed, we might now be talking about a Greater Elk Ecosystem.

This elk glyph was used by Cheyenne scout John Crazy Mule circa 1880 to indicate the Yellowstone River on a map drawn for the U.S. Cavalry.

An Unusual Hunt
Hunting is illegal in nearly all national parks, but a limited hunt occurs in Grand Teton. Congress authorized it in 1950, when the park's boundaries were expanded.

Cattle, Elk, and Disease
To keep elk numbers up, state and federal agencies in Wyoming stock select winter feeding grounds with hay. But the practice of concentrating elk is controversial. Some Yellowstone elk harbor the infectious disease brucellosis, which has been nearly eliminated from U.S. livestock. According to a federal study, these elk spread the disease to livestock 17 times between 2002 and 2012.

Lost Migrations
Parts of the Jackson Herd may once have moved south through the Green River Basin, but they no longer do. Other herds' migrations have also vanished. Scientists are working to understand exactly why. "Lost" migrations may be impossible to restore.

Map Key

Land cover: Developed; Cultivated crop; Pasture; Wetland; Water

Landownership: Wilderness; U.S. Forest Service; Private protected; National Wildlife Refuge; State park or refuge

+ Wolf reintroduction site

Elk feeding ground

Elk seasonal ranges and migration corridors
Mapped using data from select GPS-collared cow elk monitored during seasonal migration

Minor route; Major route; Winter; Summer

ALBERS CONIC EQUAL-AREA PROJECTION
SCALE 1:850,000
1 CENTIMETER = 8.5 KILOMETERS; 1 INCH = 13.4 MILES

A Visionary Proposal

Elk have always been vital to Yellowstone National Park and its adjacent lands. In 1882, U.S. Army Gen. Philip H. Sheridan was enraged to learn that as many as 4,000 elk had been slaughtered in a single winter for their hides. His prescient response was to ask that "the government extend this park ... due east about 40 miles ... [and] due south 10 miles"—an annexation that would increase "the area of the park by 3,344 square miles and would make a preserve for the large game of the West, now so rapidly decreasing." Sheridan's crusade eventually resulted in large areas abutting the park becoming protected forest reserves. The small map at right shows his proposed expansion, as well as the park's original boundary and its current one.

North American Elk Range
Elk were once ubiquitous in North America. Yet like bison, they were nearly hunted to extinction in the late 19th century. They've since rebounded and, thanks to better wildlife management, have recolonized much of their former range.

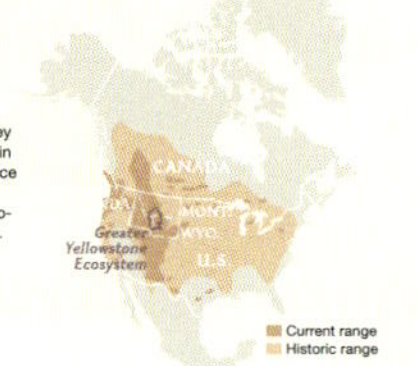

The Hunting Economy

Hunting creates a seasonal migration of its own. Each year it brings thousands of visitors to the three states—Wyoming, Montana, and Idaho—that share the ecosystem, pumping millions of dollars into their economies. If natural predators are scarce or absent, hunting can be an effective tool in limiting disease, crop losses, and unsustainable grazing. Managing elk as a game species requires officials to keep the elk population in balance with the environment—and to keep these migrations from disappearing from the landscape.

ELK HARVESTED
Share of all 2014 elk kills in the ecosystem, by hunting district

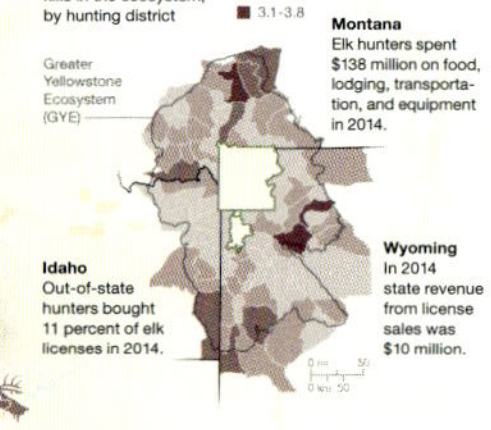

ELK HUNTING BY STATE, 2014

	IDAHO	MONTANA	WYOMING
Elk population			
STATE	105,000	160,000	114,600
Elk harvested			
GYE	2,710	7,310	10,666
STATE	20,600	25,735	25,905
Hunters			
GYE	12,866	29,446	29,957
STATE	85,000	107,663	58,266
Success rate			
GYE	21%	25%	36%
Elk hunting fee*			
RESIDENT	$43.50	$47.00	$69.50
NONRESIDENT	$571.50	$860.00**	$603.50

*Includes licenses, tags, and processing fees for 2016
**Sold only as combination license (includes other hunting, fishing)

The Comeback of the Wolf

Gray wolves were reintroduced to Yellowstone National Park starting in 1995. Since then, their numbers have risen—and their impact on the landscape and on the elk they hunt has been the subject of fierce debate. Scientists have begun to understand that the wolves' effect on the elk varies from herd to herd and year to year, depending on environmental factors such as how snow affects hunting and foraging. Biologist Doug Smith, head of the Yellowstone Wolf Project, believes the park's wolf population needs to be set by natural factors, not human management.

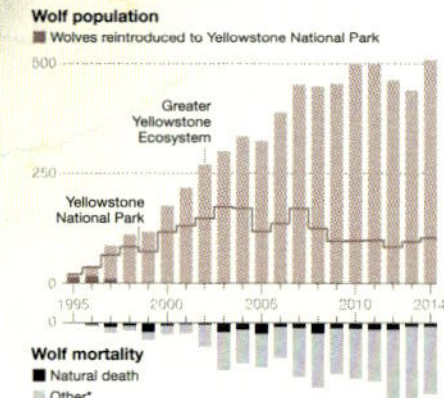

*Includes human-caused deaths and deaths from unknown causes

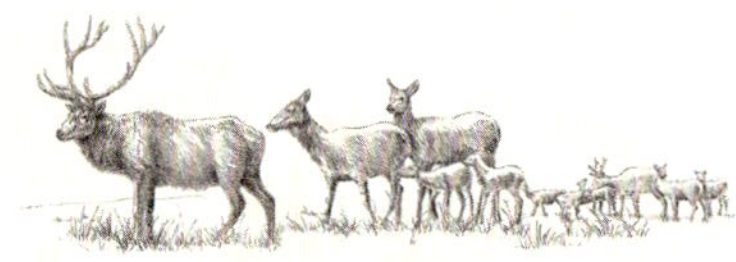

NATIONAL GEOGRAPHIC

SUPPLEMENT TO NATIONAL GEOGRAPHIC, MAY 2016
EDITOR: MARTIN GAMACHE
SCIENTIFIC ADVISER: ARTHUR MIDDLETON, YALE SCHOOL OF FORESTRY AND ENVIRONMENTAL STUDIES
GIS: THEODORE A. SICKLEY; MATTHEW M. HAYES AND JEROD A. MERKLE, WYOMING MIGRATION INITIATIVE, UNIVERSITY OF WYOMING
TEXT: JEREMY BERLIN MAP EDIT: GUS PLATIS
RESEARCH: BRAD SCRIBER, SHELLEY SPERRY
GRAPHICS: DAISY CHUNG, LAWSON PARKER
MAP PRODUCTION: DIANNE HUNT, LAUREN JAMES, LAUREN C. TIERNEY
SOURCES: MATTHEW KAUFFMAN, WYOMING MIGRATION INITIATIVE, UNIVERSITY OF WYOMING; JAMES E. MEACHAM AND ALETHEA Y. STEINGISSER, INFOGRAPHICS LAB, UNIVERSITY OF OREGON; ANDREW J. HANSEN, MONTANA STATE UNIVERSITY; SARAH DEWEY, ANN W. RODMAN, DOUGLAS W. SMITH, DANIEL STAHLER, ERIN STAHLER, P. J. WHITE, AND LEE H. WHITTLESEY, NATIONAL PARK SERVICE; ERIC COLE, NATIONAL ELK REFUGE; BRUCE B. ACKERMAN AND PAUL ATWOOD, IDAHO FISH AND GAME; RON AASHEIM, JUSTIN GUDE, KELLY PROFFITT, AND NEAL WHITNEY, MONTANA FISH, WILDLIFE AND PARKS; ALYSON COURTEMANCH, RENNY MACKAY AND DOUG MCWHIRTER, WYOMING GAME AND FISH DEPARTMENT; THE NATURE CONSERVANCY, MONTANA AND WYOMING; NATIONAL CONSERVATION EASEMENT DATABASE; PAUL C. CROSS AND LISA LANDENBURGER, NORTHERN ROCKY MOUNTAIN SCIENCE CENTER; LARRY LOENDORF, SACRED SITES

stone

esearchers
e region as
trovertial with
downers—is
tainability.
oundary
Yellowstone
ttee,

Elk River

The name Yellowstone River is likely derived from a term used by the Hidatsa tribe. But in the Crow, Lakota, and Cheyenne languages, it was called Elk River. The importance of the animals to these last three tribes can be found in petroglyph sites, antler monuments, and ritual dress. If their name for the river had prevailed, we might now be talking about a Greater Elk Ecosystem.

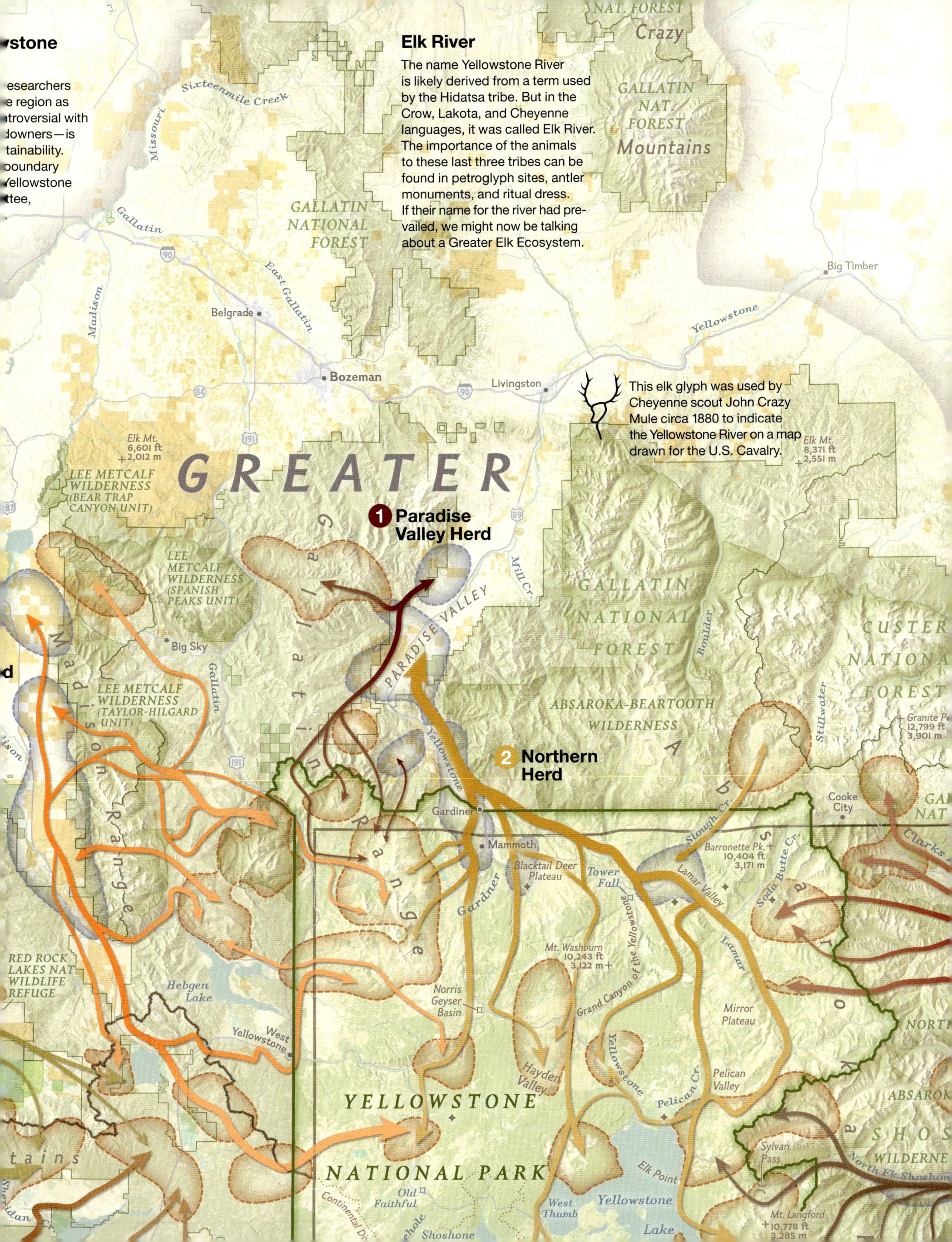

This elk glyph was used by Cheyenne scout John Crazy Mule circa 1880 to indicate the Yellowstone River on a map drawn for the U.S. Cavalry.

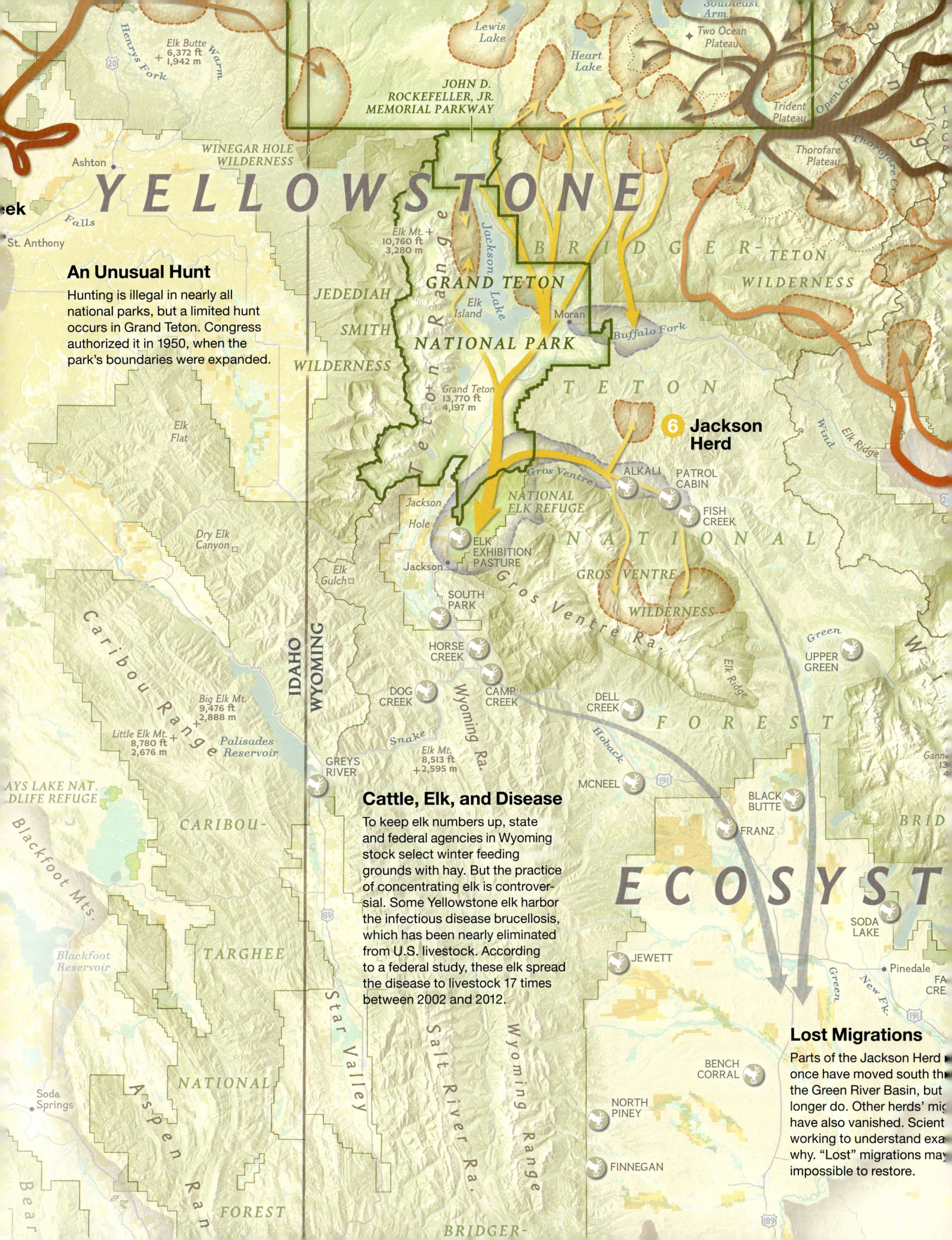

An Unusual Hunt

Hunting is illegal in nearly all national parks, but a limited hunt occurs in Grand Teton. Congress authorized it in 1950, when the park's boundaries were expanded.

Cattle, Elk, and Disease

To keep elk numbers up, state and federal agencies in Wyoming stock select winter feeding grounds with hay. But the practice of concentrating elk is controversial. Some Yellowstone elk harbor the infectious disease brucellosis, which has been nearly eliminated from U.S. livestock. According to a federal study, these elk spread the disease to livestock 17 times between 2002 and 2012.

Lost Migrations

Parts of the Jackson Herd once have moved south th[e] the Green River Basin, but longer do. Other herds' mig[rations] have also vanished. Scient[ists are] working to understand exa[ctly] why. "Lost" migrations may [be] impossible to restore.

1:5,500,000

ANTARCTICA

Polar Geospatial Center

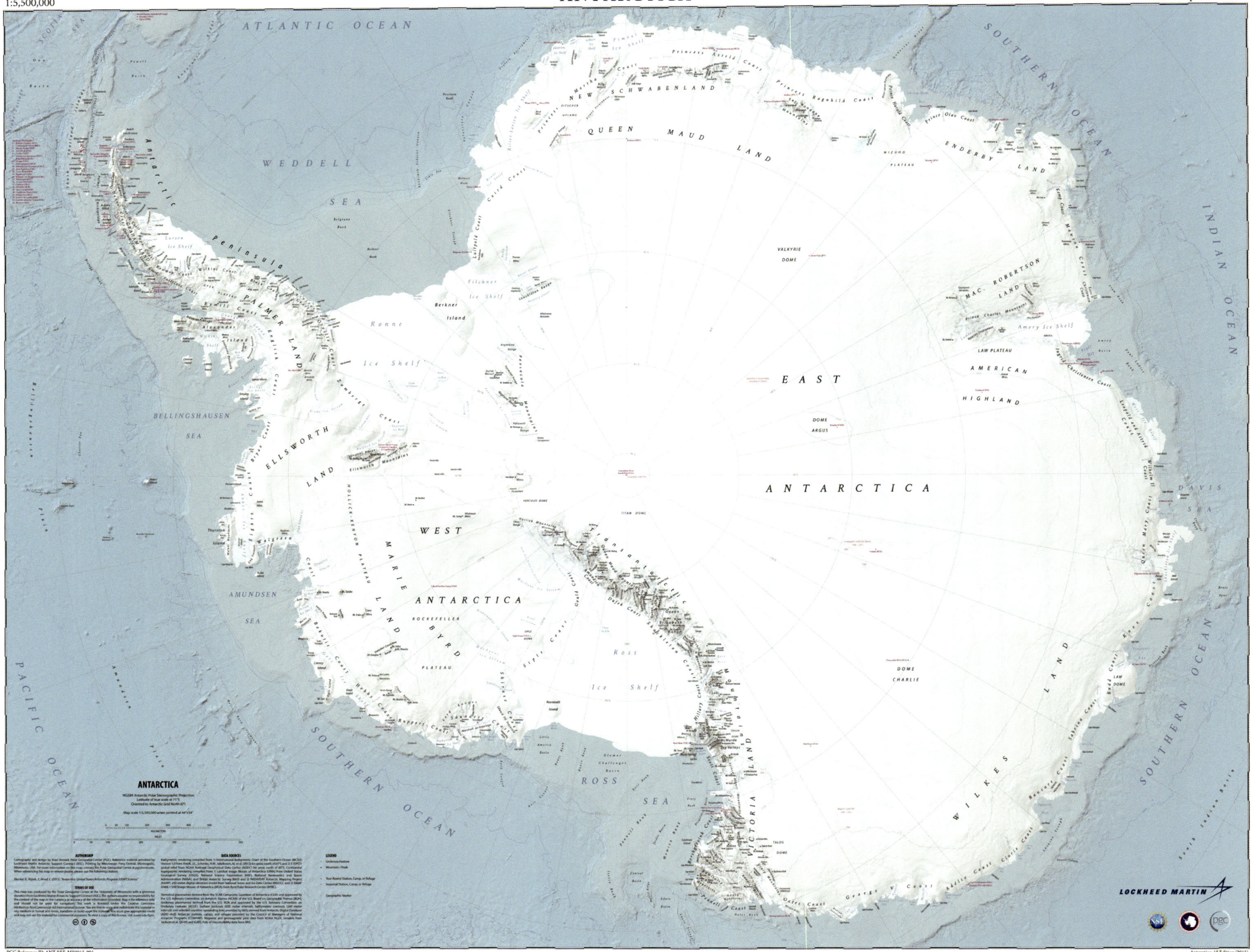

Antarctica: 1st Edition (2015)

《 Antarctica

Brad Herried | Polar Geospatial Center

From the 18th century to the 21st, in the popular imagination Antarctica has been terra incognita, an unknown territory, a mysterious land of ice and penguins. In reality, the continent is home to a variety of stunning landscapes: the high, jagged peaks of the Transantarctic Mountains; the surf-bashed, rocky beaches of the Antarctic Peninsula islands; the isolated, moon-like ecosystem of the McMurdo Dry Valleys; and the windswept sastrugi atop the world's largest ice cap.

This map of Antarctica and the accompanying scientific poster on the reverse capture the variety of Antarctic geography and educate readers about the continent's wonders and ongoing scientific projects. Our goal was a comprehensive, educational reference map that would be useful for teachers, planners, researchers, and the general public. With this map, we visually and spatially contextualize Antarctica, depicting the vast natural landscape, highlighting locations that support human presence for research, and showing just how much of the continent is yet to be explored and understood. We note where scientific stations are located and which geographic features have been named. Antarctica has no road networks, no official administrative boundaries, no capital cities. Maps must therefore reflect unique, remote geographies, representing the place as the sum of its magnificent, unexplored parts.

Today the map hangs in hundreds of classrooms, research labs, and government agencies across the United States. A generous donation from Lockheed Martin, primary logistical contractor to the National Science Foundation (NSF) U.S. Antarctic Program, to the Polar Geospatial Center, an NSF geospatial research support center at the University of Minnesota, supported the creation of the map. Free access to the digital map and science poster is available at z.umn.edu/antmap.

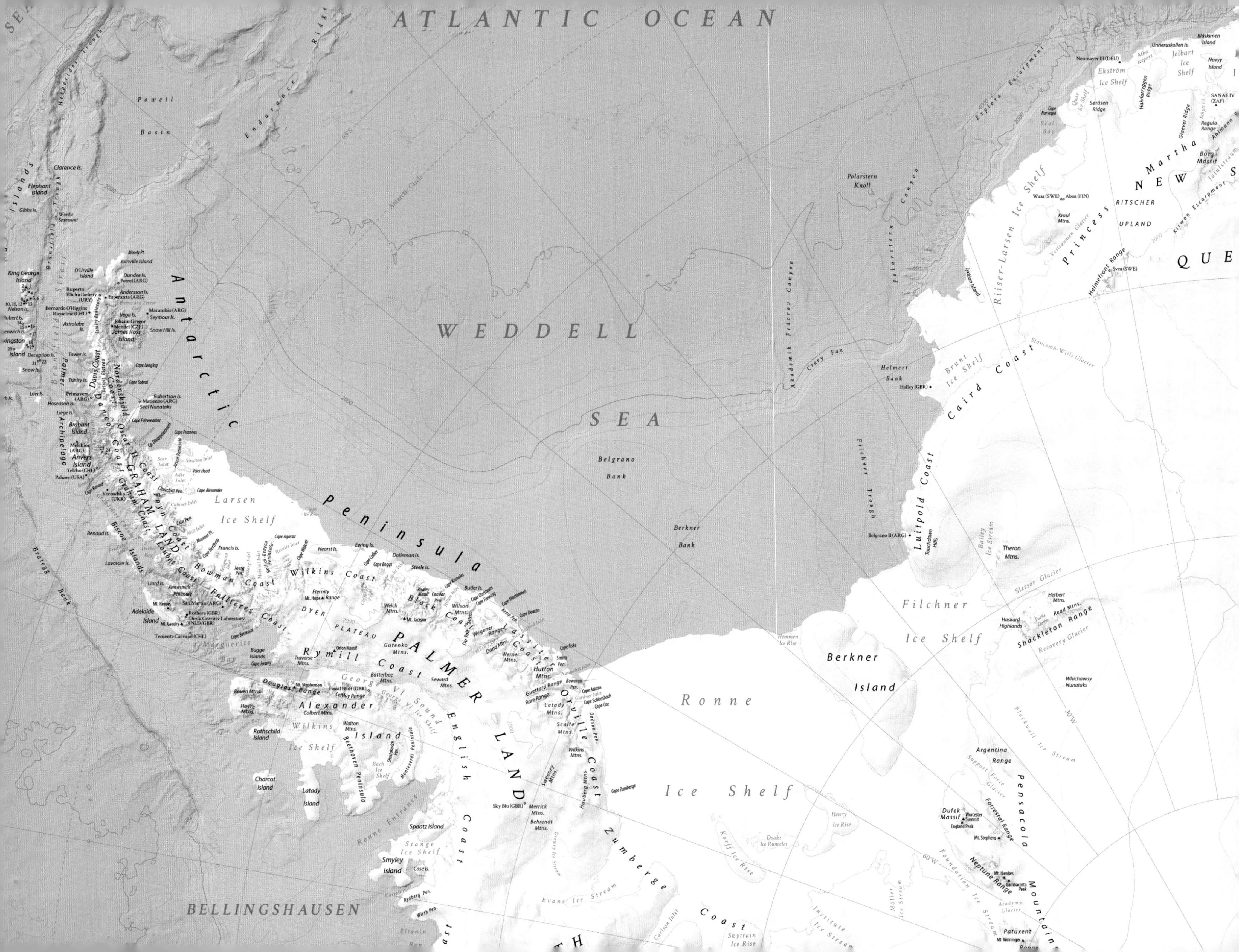
ATLANTIC OCEAN
Powell Basin
Endurance Ridge
Antarctic Circle
Clarence Is.
Elephant Island
Gibbs Is.
Wordie Seamount
Bransfield Trough
Hesperides Trough
Moody Pt.
Joinville Island
D'Urville Island
Dundee Is.
Petrel (ARG)
King George Island
Ruperto Elichiribehety (URY)
Esperanza (ARG)
Andersson Is.
Bernardo O'Higgins
Riquelme (CHL)
Marambio (ARG)
Seymour Is.
Vega Is.
Johann Gregor Mendel (CZE)
James Ross Island
Snow Hill Is.
Nelson Is.
Astrolabe Is.
Tower Is.
Deception Is.
Snow Is.
Cape Longing
Cape Sobral
Trinity Is.
Low Is.
Primavera (ARG)
Hoseason Is.
Robertson Is.
Matienzo (ARG)
Seal Nunataks
Liege Is.
Brabant Island
Cape Fairweather
Palmer Archipelago
Melchior (ARG)
Anvers Island
Yelcho (CHL)
Palmer (USA)
Vernadsky (UKR)
Cape Framnes
Scar Inlet
Stratton Inlet
Veier Head
Churchill Pen.
Cape Alexander
Cabinet Inlet
Antarctic Peninsula
Davis Coast
Danco Coast
Nordenskjöld Coast
Oscar II Coast
Graham Coast
Foyn Coast
Bowman Coast
Loubet Coast
Fallières Coast
GRAHAM LAND
Larsen Ice Shelf
Biscoe Islands
Renaud Is.
Lavoisier Is.
Francis Is.
Cape Agassiz
Hearst Is.
Ewing Is.
Dolleman Is.
Steele Is.
Wilkins Coast
Black Coast
Lassiter Coast
Butler Is.
Cape Knowles
Cape Mackintosh
Kemp Pen.
Cape Deacon
Eternity Range
Mt. Hope
Welch Mtns.
Wilson Mtns.
Mt. Jackson
Adelaide Island
San Martín (ARG)
Rothera (GBR)
Dirck Gerritsz Laboratory (NLD/GBR)
Teniente Carvajal (CHL)
Marguerite Bay
Bugge Islands
Cape Jeremy
DYER PLATEAU
Orion Massif
Gutenko Mtns.
Traverse Mtns.
Rymill Coast
PALMER LAND
Batterbee Mtns.
Seward Mtns.
Dana Mtns.
Werner Mtns.
Hutton Mtns.
Cape Fiske
Smith Pen.
Bowman Pen.
Cape Adams
Cape Schlossbach
Cape Cox
Douglas Range
Mt. Stephenson
Fossil Bluff (GBR)
LeMay Range
George VI Sound
Alexander Island
Colbert Mtns.
Rouen Mtns.
Havre Mtns.
Rothschild Island
Wilkins Ice Shelf
Walton Mtns.
Beethoven Peninsula
Bach Ice Shelf
Guettard Range
Rare Range
Latady Mtns.
Scaife Mtns.
Wilkins Mtns.
Orville Coast
English Coast
Charcot Island
Latady Island
Sweeney Mtns.
Hauberg Mtns.
Cape Zumberge
Sky Blu (GBR)
Merrick Mtns.
Behrendt Mtns.
Ronne Entrance
Spaatz Island
Stange Ice Shelf
Smyley Island
Case Is.
Rydberg Pen.
Wirth Pen.
Evans Ice Stream
Zumberge Coast
Carlson Inlet
Skytrain Ice Rise
Institute Ice Stream
Möller Ice Stream
Foundation Ice Stream
BELLINGSHAUSEN
Bryant Bank
WEDDELL SEA
Belgrano Bank
Berkner Bank
Hemmen Ice Rise
Berkner Island
Ronne Ice Shelf
Henry Ice Rise
Korff Ice Rise
Doake Ice Rumples
Filchner Ice Shelf
Filchner Trough
Polarstern Knoll
Polarstern Canyon
Akademik Fedorov Canyon
Crary Fan
Helmert Bank
Halley (GBR)
Brunt Ice Shelf
Caird Coast
Stancomb-Wills Glacier
Luitpold Coast
Belgrano II (ARG)
Touchdown Hills
Bailey Ice Stream
Theron Mtns.
Slessor Glacier
Herbert Mtns.
Read Mtns.
Haskard Highlands
Shackleton Range
Recovery Glacier
Whichaway Nunataks
Blackwall Ice Stream
Argentina Range
Support Force Glacier
Forrestal Range
Dufek Massif
Worcester Summit
England Peak
Mt. Stephens
Pensacola Mountains
Neptune Range
Mt. Hawkes
Gambacorta Peak
Academy Glacier
Patuxent Range
Mt. Weininger
Explora Escarpment
Neumayer III (DEU)
Atka Iceport
Ekström Ice Shelf
Jelbart Ice Shelf
Unneruskollen Is.
Blåskimen Island
Novyy Island
Quar Ice Shelf
Halvfarryggen Ridge
Sørdsen Ridge
Cape Norvegia
Seal Bay
SANAE IV (ZAF)
Giaever Ridge
Regula Range
Ahlmann Ridge
Borg Massif
Riiser-Larsen Ice Shelf
Wasa (SWE)
Aboa (FIN)
Kraul Mtns.
Vestraumen Glacier
Princess Martha
NEW S
RITSCHER UPLAND
Kirwan Escarpment
Heimefront Range
Svea (SWE)
Lyddan Island
QUE

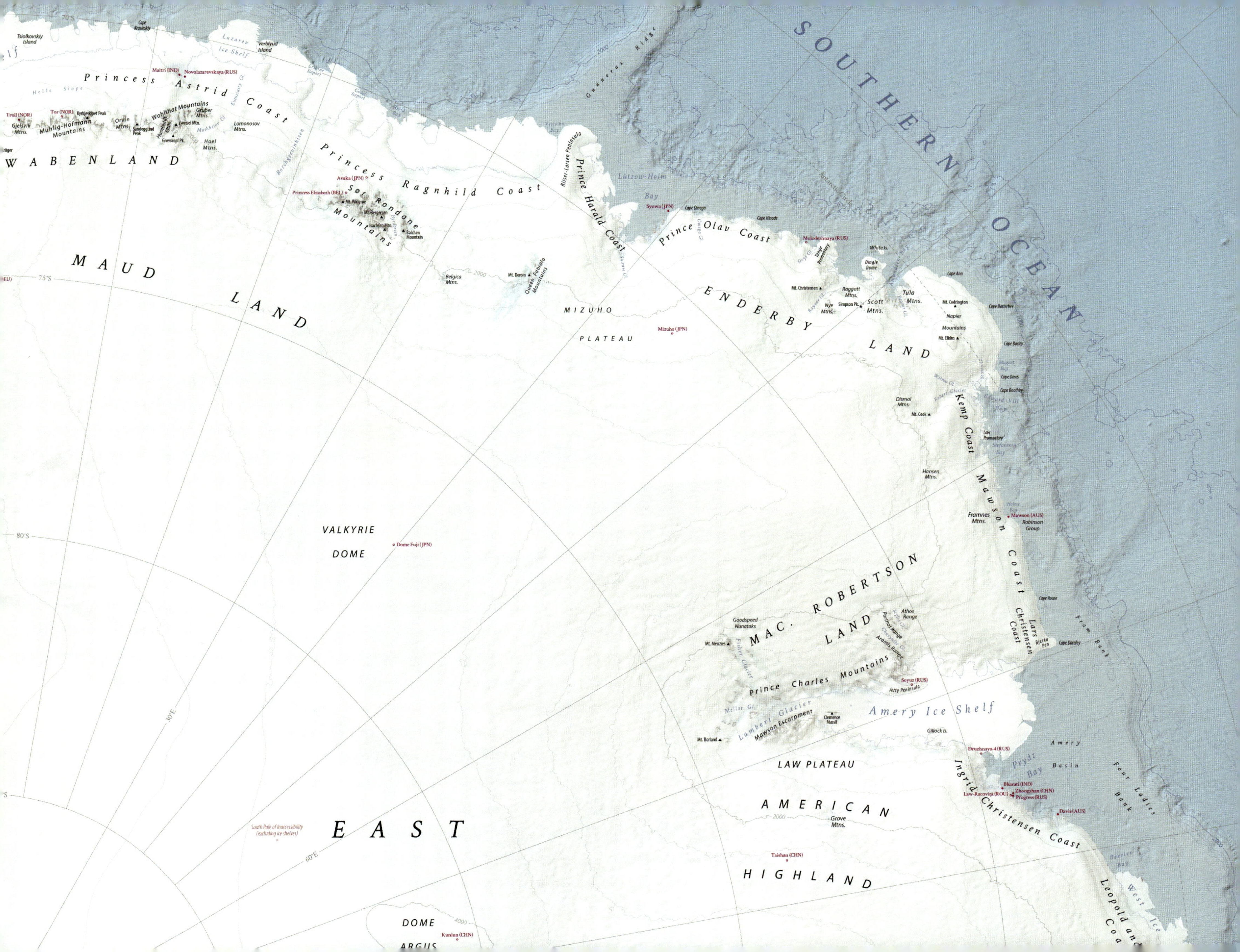

SOUTHERN OCEAN
Tsiolkovskiy Island
Cape Kosoaksky
Lazarev Ice Shelf
Verblyud Island
Maitri (IND)
Novolazarevskaya (RUS)
Princess Astrid Coast
Helle Slope
Troll (NOR)
Tor (NOR)
Gjelsvik Mtns.
Mühlig-Hofmann Mountains
Orvin Mtns.
Wohlthat Mountains
Gruber Mtns.
Humboldt Mtns.
Lomonosov Mtns.
Hoel Mtns.
WABENLAND
Gunnerus Ridge
Princess Ragnhild Coast
Asuka (JPN)
Princess Elisabeth (BEL)
Sør Rondane Mountains
Balchen Mountain
Riiser-Larsen Peninsula
Prince Harald Coast
Lützow-Holm Bay
Syowa (JPN)
Cape Omega
Prince Olav Coast
Cape Hinode
Molodezhnaya (RUS)
MAUD LAND
Belgica Mtns.
Queen Fabiola Mountains
MIZUHO PLATEAU
Mizuho (JPN)
ENDERBY LAND
Mt. Christensen
Raggatt Mtns.
Nye Mtns.
Scott Mtns.
Tula Mtns.
Napier Mountains
Cape Ann
Cape Batterbee
Cape Borley
Cape Davis
Cape Boothby
Edward VIII Bay
Kemp Coast
Dismal Mtns.
Mt. Cook
Hansen Mtns.
Mawson Coast
Framnes Mtns.
Mawson (AUS)
Robinson Group
VALKYRIE DOME
Dome Fuji (JPN)
MAC. ROBERTSON LAND
Goodspeed Nunataks
Athos Range
Porthos Range
Aramis Range
Mt. Menzies
Fisher Glacier
Lars Christensen Coast
Cape Darnley
Fram Bank
Prince Charles Mountains
Soyuz (RUS)
Jetty Peninsula
Amery Ice Shelf
Mellor Gl.
Lambert Glacier
Mawson Escarpment
Clemence Massif
Mt. Borland
Gillock Is.
Druzhnaya-4 (RUS)
LAW PLATEAU
Prydz Bay
Amery Basin
Four Ladies Bank
Bharati (IND)
Zhongshan (CHN)
Law-Racoviță (ROU)
Progress (RUS)
Davis (AUS)
Ingrid Christensen Coast
AMERICAN HIGHLAND
Grove Mtns.
South Pole of Inaccessibility (excluding ice shelves)
EAST
Taishan (CHN)
Barrier Bay
West Ice
Leopold and Coa
DOME ARGUS
Kunlun (CHN)
75°S
80°S
30°E
60°E

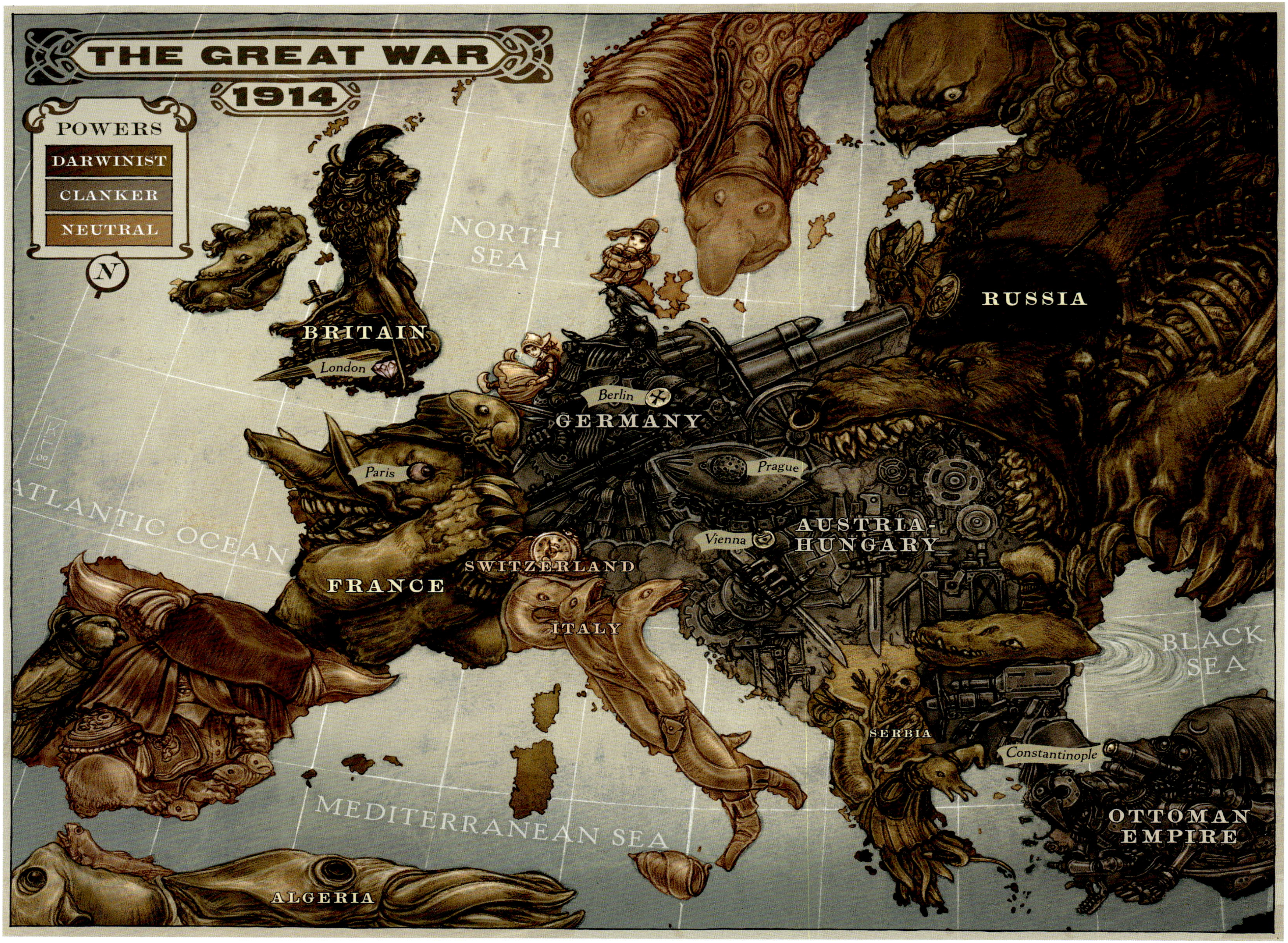
THE GREAT WAR
1914
POWERS
DARWINIST
CLANKER
NEUTRAL
N
NORTH SEA
BRITAIN
London
RUSSIA
Berlin
GERMANY
Paris
Prague
ATLANTIC OCEAN
AUSTRIA-HUNGARY
Vienna
SWITZERLAND
FRANCE
ITALY
BLACK SEA
SERBIA
Constantinople
OTTOMAN EMPIRE
MEDITERRANEAN SEA
ALGERIA

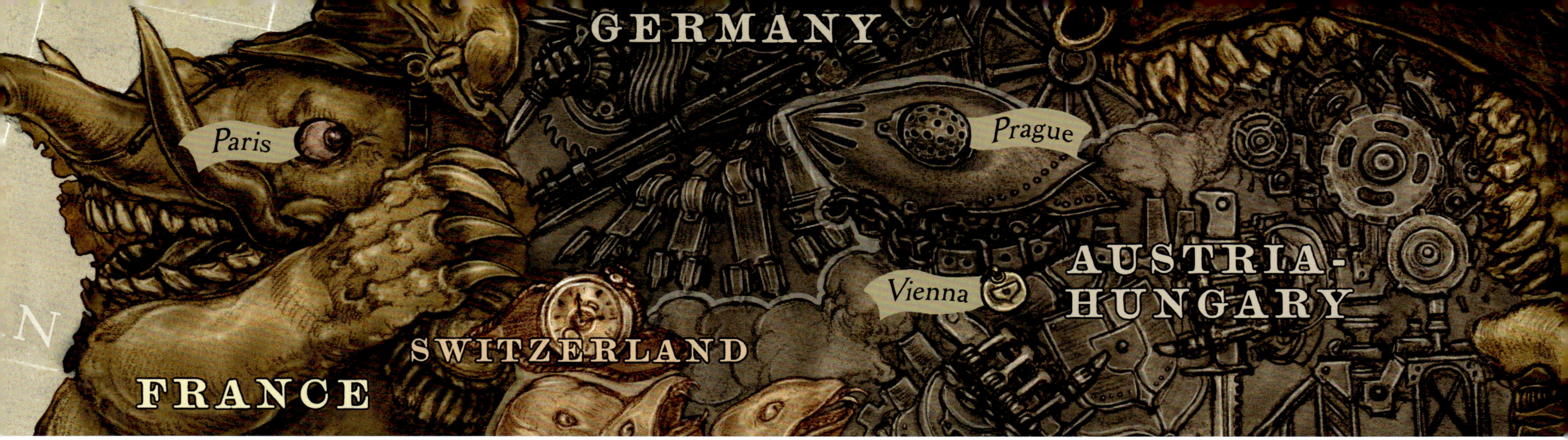

《 Caricature Map of Europe 1914

Keith Thompson

This is a political caricature map depicting European nations as described in the alternate world history of *Leviathan*, a novel by Scott Westerfeld. In the novel, the Clanker Powers of Germany, Austria-Hungary, and Turkey are pitted against the Darwinist Powers of Great Britain, France, Russia, and Serbia. The countries are caricatured based on the perspective of a British publication in the early twentieth century.

Interior illustration from Leviathan by Scott Westerfeld illustrated by Keith Thompson (keiththompsonart.com), from Simon Pulse, an imprint of Simon & Schuster Children's Publishing Division

The Clanker Powers: Germany is a massive military machine with weapons aimed outwards toward all surrounding countries. Germany's weapons point at Great Britain—less as a sign of direct aggression and more as an indicator that Germany is on the verge of starting a grand global Empire to challenge the current one. Austria-Hungary is an aggressive armoured giant, teetering on shoddy foundations. It is also the primary aggressor in a land grab against Serbia, with two bayonets piercing the border. The Ottoman Empire is a teetering automaton, collapsing under the weight of a paranoid and ungainly spy network that peers across Europe through multiple lenses and spyglasses.

The Darwinist Powers: Great Britain is a lion with a Roman Imperial helmet, sitting on a mound of riches gathered from its Empire. France's elephant beast wears the French kepi soldiers started the Great War with before adopting firefighter helmets. The image of France is influenced by the Elephantine Colossus built for the Universal Exhibition of 1889 in Paris. Russia is a huge imperialist bear, rotting and filled with maggots. The skeletal imagery representing Serbia indicates the huge numbers of civilian deaths the nation suffered in the war. Portugal is a parrot for the Entente trying to goad a slumbering Spain into action. Ireland looks askance at Britain and brandishes a shillelagh, an indicator of a rough relationship, and Ireland's eventual involvement with the Clanker Powers.

Neutral Nations: Norway and Sweden are both Scandinavian trolls in the style of John Bauer, an inspirational illustrator from the era who produced phenomenal work during the war. Italy is a clutch of snakes intent on joining the Clanker Powers, despite existing agreements. The neutral Swiss watch ticks away the time, comfortable to wait it all out.

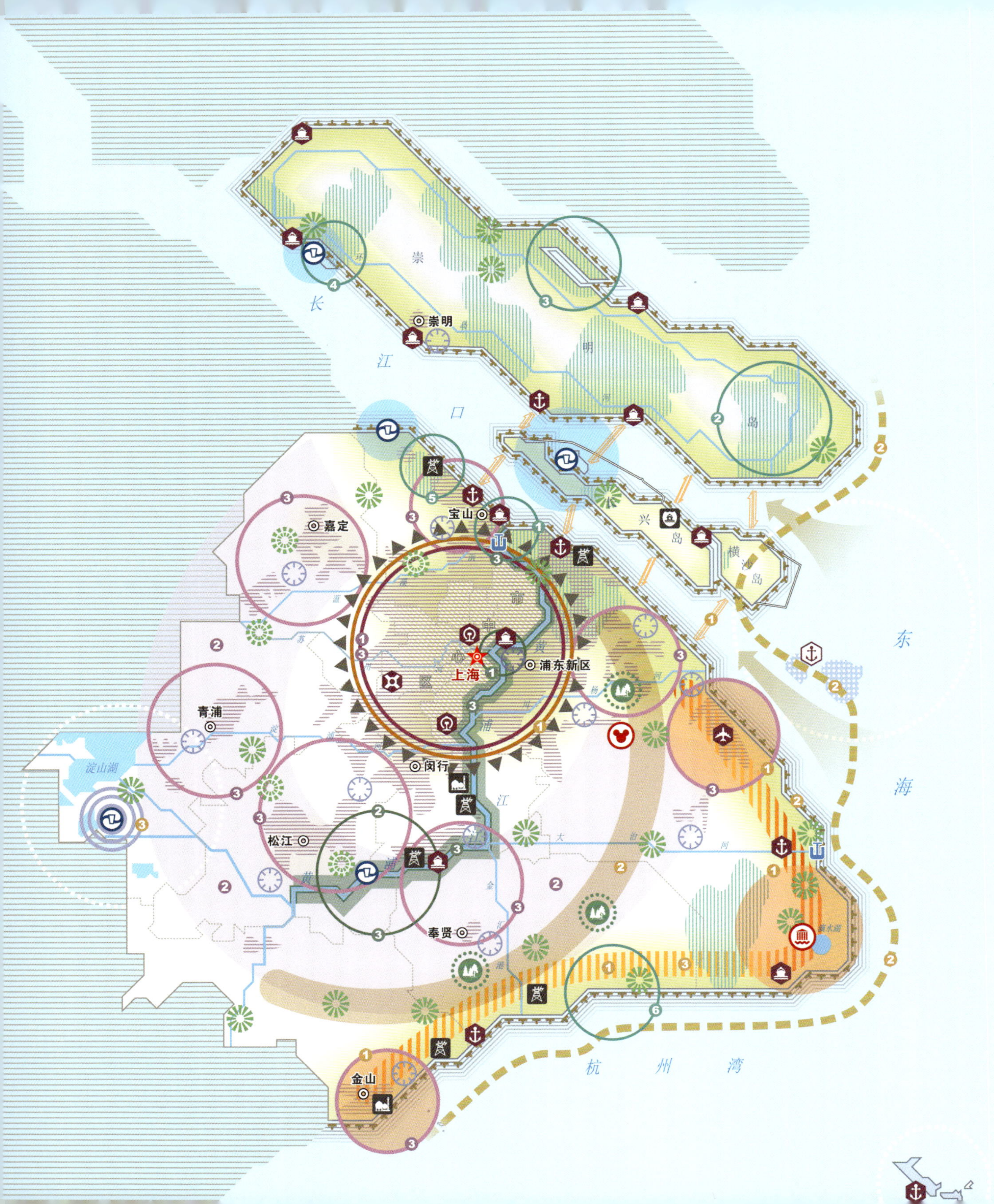

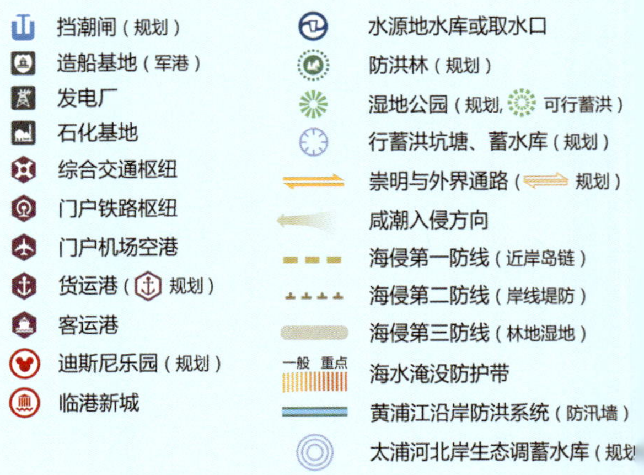

反‘t’型格局

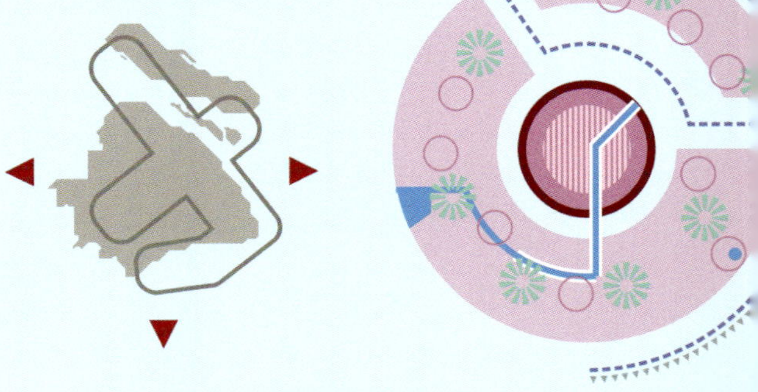

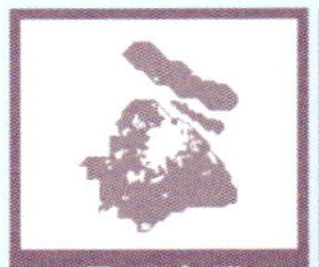

暴雨内涝

1. 市区“再自然化”措施：增加绿化屋顶、透水路面、道路绿
排水系统，建立地上/地下贮水池等人工水体，恢复河网、
能公园区（平时为休憩景观区，战时为行蓄洪区）等，增加
2. 郊区“生态保育”措施：增建湿地林地，减少降雨产、汇
系密布，骨干河道定期疏通清障，保障防洪、蓄洪能力；
3. 防涝标准提升：市区及周边扩张区50-100年一遇（主干
浦江上游地区）30-50年一遇；郊区其他区域10-30年一

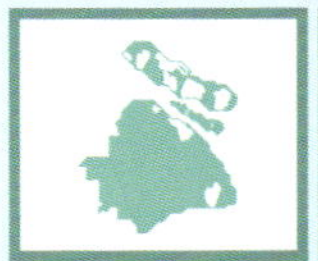

台风风暴潮

1. 黄浦江河口新建挡潮闸，吴淞口附近低洼区域可开辟分洪区。
2. 崇明东部向化镇、中兴镇、陈家镇等人口集中区域应考虑建
3. 崇明北沿海塘防台抗洪能力较弱，尤其是兴隆沙附近，至204
4. 崇明南沿部分海塘应新建或加固，尤其是崇头、新河、堡镇、
5. 宝山海塘应重视地面沉降及海平面上升影响，至2040年应加
6. 杭州湾北岸属冲刷岸段，应加强保滩工程措施，尤其是芦潮

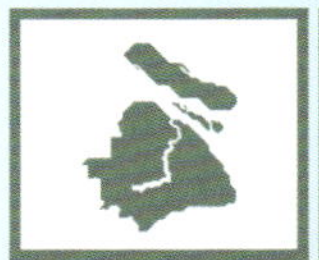

黄浦江风暴潮

1. 在外滩、陆家嘴等重点区域建设地下蓄水设施或改造利用防
2. 在中上游地区，尤其松江、闵行段等低洼区域，开辟数个分
3. 建议一：黄浦江河口建闸，防止水位过快上涨，最大程度地
建议二：加固中上游薄弱段和新增城市化地区防洪系统，工
上游、中下游防汛墙需维持或加高提升至千年一遇

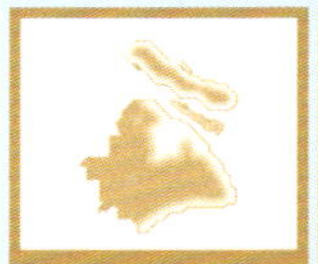

海平面上升淹没

1. 半径12 公里市中心城区为高风险核心保护区；沿海14公里
设施，已有重要设施需重点保护；加强崇明岛对外交通联系，
2. 三道防线：第一道由滩涂-湿地-岛屿组成（加强保护崇明东
（提高堤防设计标准），第三道由现有及规划的郊野湿地公园
3. 建议一：通过海滩养护和防波堤建设，对沿海近岸岛屿、滩
建议二：落实黄浦江上游生态调蓄水库的建设，削减咸潮入
建议三：沿海地区应划定10公里范围，严格控制高层建筑，

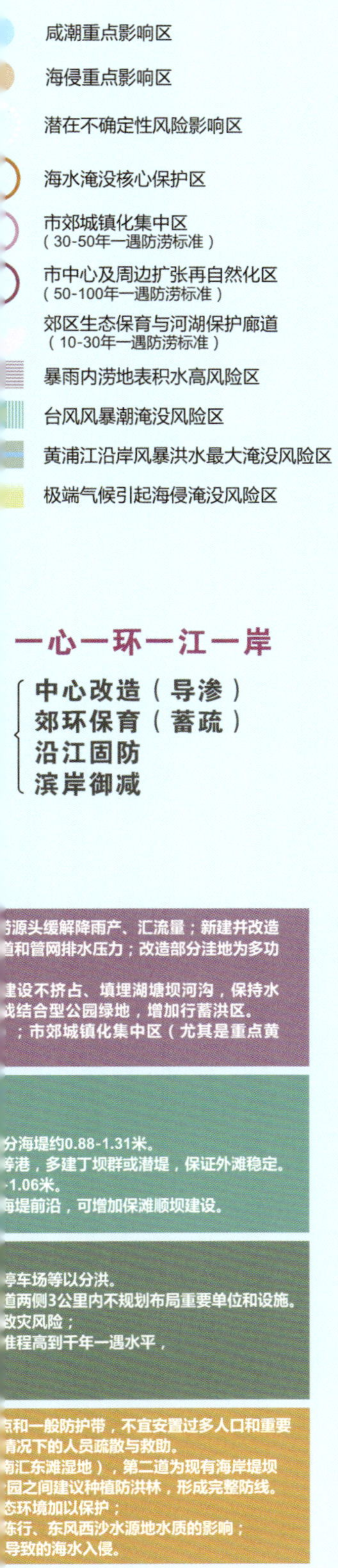

上海市综合防灾空间策略概念图景

« Alternative Strategic Spatial Vision of Natural Disaster Prevention in Future Shanghai

唐曦, *Xi TANG* | *East China Normal University*

This map serves as the key cartographic diagram in a final report for the project, "Comprehensive Disaster Prevention and Urban Safety Systems in Shanghai," part of a larger research effort, Strategic Studies on a New Round of Shanghai Urban Master Planning. In this map, I describe alternative spatial strategies responding to probable natural hazards that will affect Shanghai in the future.

Showing diverse disaster scenarios in such a limited physical mapping space is a challenge because of the complex overlapping of point, line, and area objects. Information overload for viewers was the primary problem I had to solve. Another vital task was to communicate the thematic content clearly to an uninitiated audience and a variety of stakeholders.

I focused on harmonizing the relationship between complexity and usability. During the design stage, I used effective symbol generalization, self-metaphorical visual logic, and rational formal integration to construct a storytelling stage with explicit visual hierarchies. The base map was extremely simplified, retaining just the main river network, administrative division boundaries, and shorelines—the reference elements most closely related to the map's themes. I used schematic representation in keeping with the classic tube-map style for line objects. I also used a four-color category system to distinguish four thematic groups, including rainstorm waterlogging (purple), typhoon storm surge (teal green), Huangpu River storm surge (dark ocean green), and extreme sea-level rise (orange). A light sky-blue background helps to convey the setting of the ocean city.

In order to minimize interference over the map, I used transparency and varied texture to allow elements to coexist despite overlapping, and to allow for adaptation to future blueprints. A horizontal split layout of integrated graphical representations reduced map information density: the main map is on the left, with the legend and accompanying explanatory diagrams on the right. Two abstract schematics related to the core planning concepts are included to assist in grasping the patterns in a macro perspective.

On the whole, the map design implements an integrative geo-infographic design, which no longer adheres rigidly to strict cartographic mathematical bases, but still keeps the communication of information as the primary goal. This methodology of mutual reference and superior reciprocity among infographics, thematic maps, and planning illustrations, could optimize future spatial planning visualizations through a creative and aesthetically pleasing style.

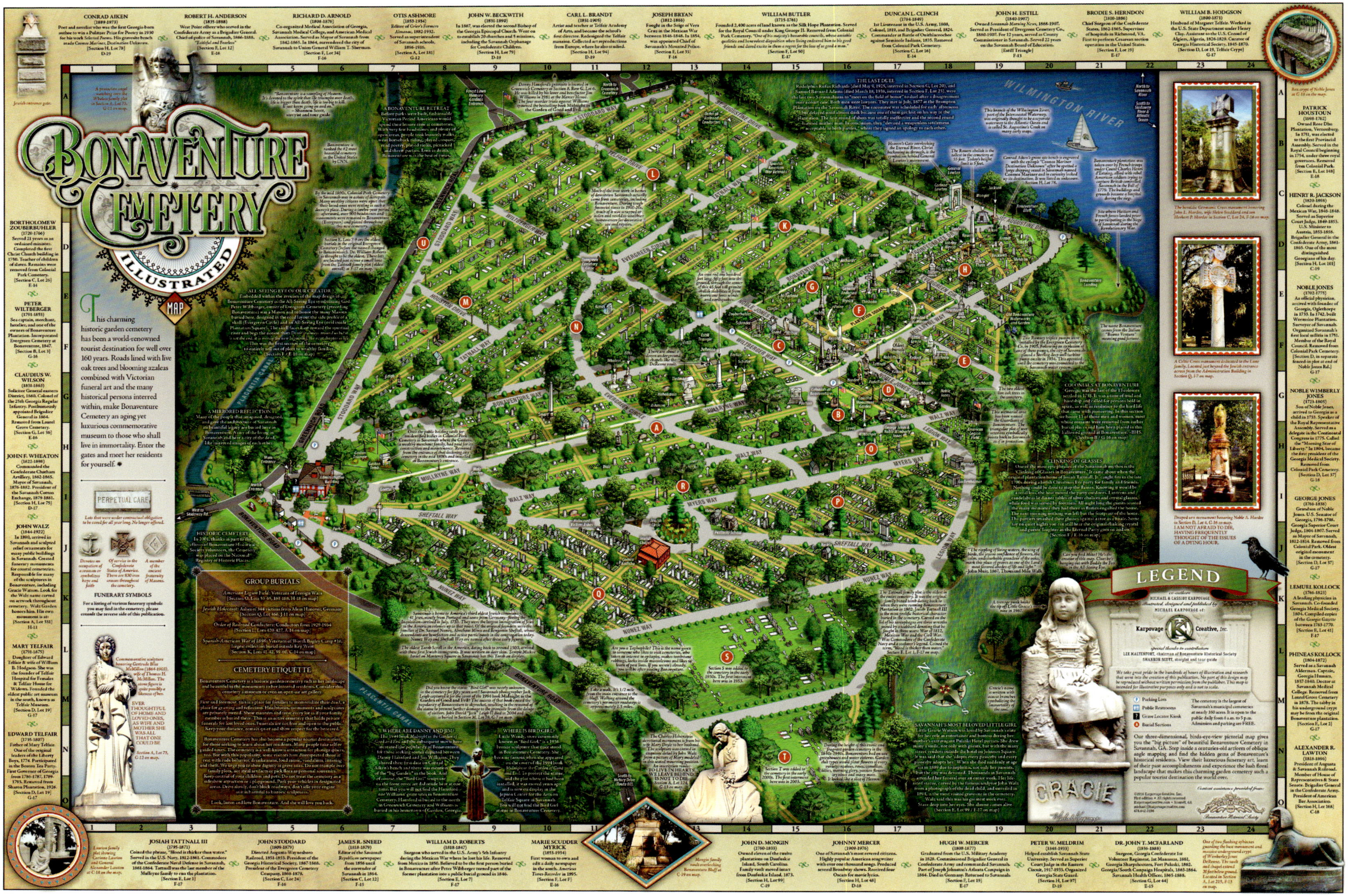
BONAVENTURE CEMETERY
ILLUSTRATED
MAP
This charming historic garden cemetery has been a world-renowned tourist destination for well over 160 years. Roads lined with live oak trees and blooming azaleas combined with Victorian funeral art and the many historical persons interred within, make Bonaventure Cemetery an aging yet luxurious commemorative museum to those who shall live in immortality. Enter the gates and meet her residents for yourself.
PERPETUAL CARE
Lots that were under contractual obligation to be cared for all year long. No longer offered.
FUNERARY SYMBOLS
GROUP BURIALS
CEMETERY ETIQUETTE
WILMINGTON RIVER
LEGEND
GRACIE
CONRAD AIKEN
ROBERT H. ANDERSON
RICHARD D. ARNOLD
OTIS ASHMORE
JOHN W. BECKWITH
CARL L. BRANDT
JOSEPH BRYAN
WILLIAM BUTLER
DUNCAN L. CLINCH
JOHN H. ESTILL
BRODIE S. HERNDON
WILLIAM B. HODGSON
PATRICK HOUSTOUN
HENRY R. JACKSON
NOBLE JONES
NOBLE WIMBERLY JONES
GEORGE JONES
LEMUEL KOLLOCK
PHINEAS KOLLOCK
ALEXANDER R. LAWTON
BORTHOLOMEW ZOUBERBUHLER
PETER WILTBERGER
CLAUDIUS W. WILSON
JOHN F. WHEATON
JOHN WALZ
MARY TELFAIR
EDWARD TELFAIR
JOSIAH TATTNALL III
JOHN STODDARD
JAMES R. SNEED
WILLIAM D. ROBERTS
MARIE SCUDDER MYRICK
JOHN D. MONGIN
JOHNNY MERCER
HUGH W. MERCER
PETER W. MELDRIM
DR. JOHN T. MCFARLAND

« Bonaventure Cemetery Illustrated Map

Michael Karpovage

Bonaventure Cemetery is one of Savannah, Georgia's most visited destinations for tourists and locals alike. Getting lost in the cemetery's 100 acres and thousands of monuments was almost guaranteed before we created the "Bonaventure Cemetery Illustrated Map." Now visitors to this haunting, historic garden cemetery can see an initial big picture of its crisscrossing avenues, aisles, and alleys, before finding its most famous residents. When you unfold the poster-size map, you'll see illustrations highlighting Bonaventure's most visited graves.

I made this 3-D, birds-eye-view pictorial map for taphophiles—people who visit cemeteries to learn the history of those buried there and who appreciate the monuments, symbolism, and funerary art of these outdoor museums.

The map's artwork is reminiscent of 19th-century cartography, fitting for a cemetery established in the Victorian era. After visiting Bonaventure several times as a tourist and getting lost myself, I thought the public deserved an upgrade in navigational tools. As a Savannah history lover, I also knew that Bonaventure itself deserved a beautiful map.

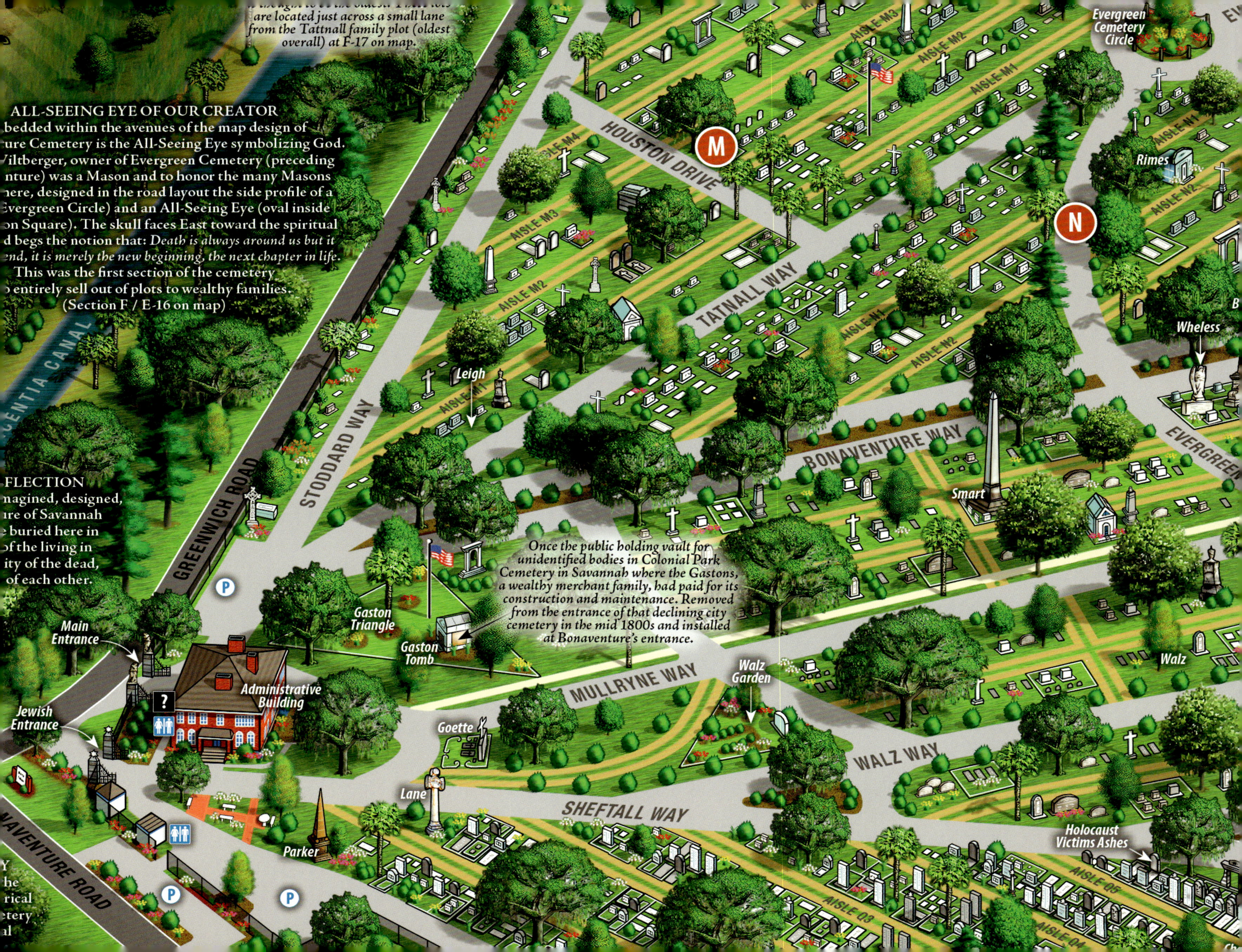
are located just across a small lane
from the Tattnall family plot (oldest
overall) at F-17 on map.
ALL-SEEING EYE OF OUR CREATOR
bedded within the avenues of the map design of
ure Cemetery is the All-Seeing Eye symbolizing God.
iltberger, owner of Evergreen Cemetery (preceding
nture) was a Mason and to honor the many Masons
here, designed in the road layout the side profile of a
vergreen Circle) and an All-Seeing Eye (oval inside
on Square). The skull faces East toward the spiritual
d begs the notion that: *Death is always around us but it*
end, it is merely the new beginning, the next chapter in life.
This was the first section of the cemetery
entirely sell out of plots to wealthy families.
(Section F / E-16 on map)
FLECTION
nagined, designed,
ire of Savannah
buried here in
f the living in
ity of the dead,
of each other.
Once the public holding vault for
unidentified bodies in Colonial Park
Cemetery in Savannah where the Gastons,
a wealthy merchant family, had paid for its
construction and maintenance. Removed
from the entrance of that declining city
cemetery in the mid 1800s and installed
at Bonaventure's entrance.
Evergreen Cemetery Circle
M
N
HOUSTON DRIVE
TATTNALL WAY
BONAVENTURE WAY
MULLRYNE WAY
WALZ WAY
SHEFTALL WAY
STODDARD WAY
GREENWICH ROAD
AVENTURE ROAD
ENTIA CANAL
EVERGREEN
AISLE M1
AISLE M2
AISLE M3
AISLE M4
AISLE N1
AISLE N2
AISLE O3
AISLE O4
AISLE O5
Rimes
Wheless
Leigh
Smart
Walz
Walz Garden
Gaston Triangle
Gaston Tomb
Goette
Lane
Parker
Main Entrance
Administrative Building
Jewish Entrance
Holocaust Victims Ashes
P

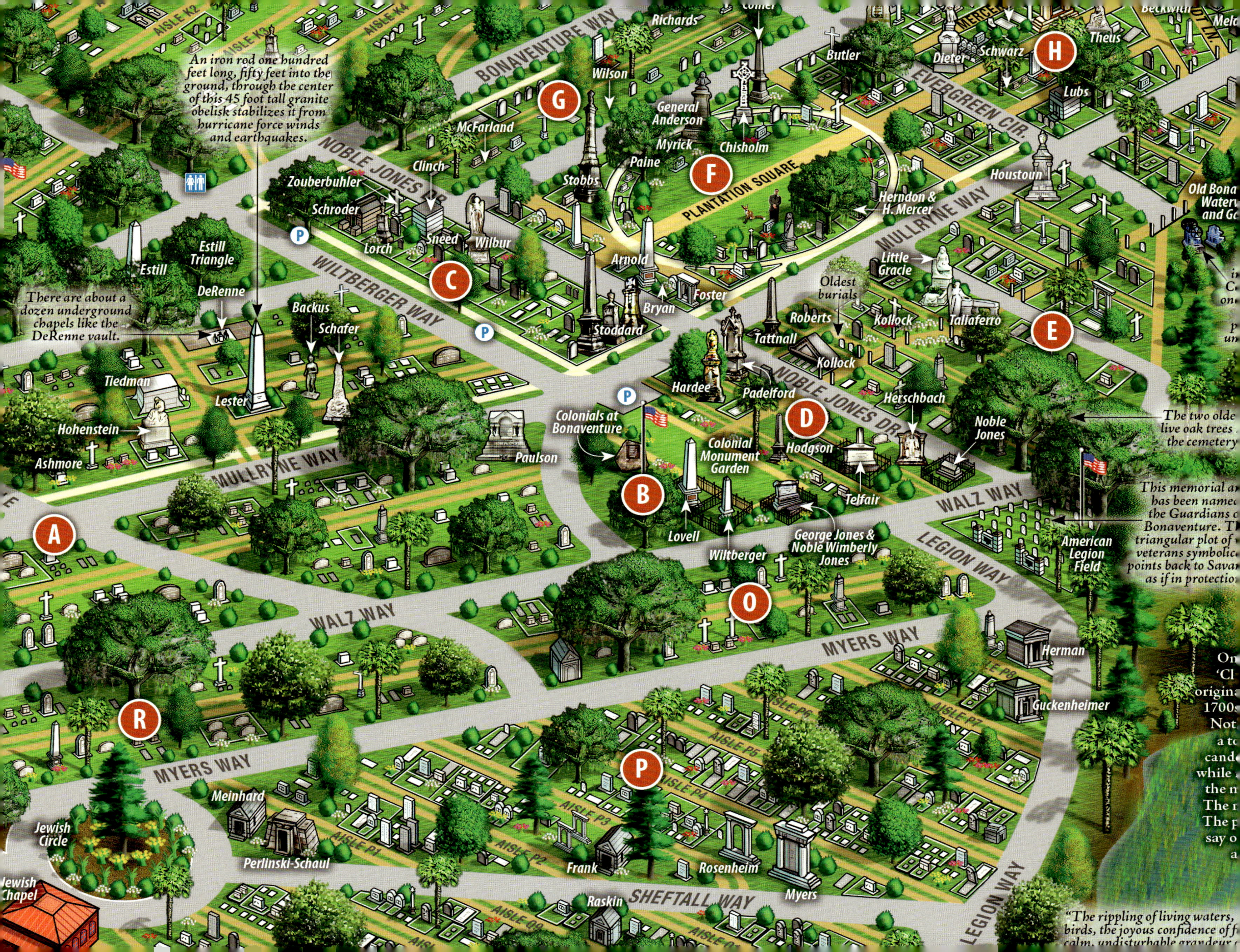

An iron rod one hundred feet long, fifty feet into the ground, through the center of this 45 foot tall granite obelisk stabilizes it from hurricane force winds and earthquakes.
There are about a dozen underground chapels like the DeRenne vault.
AISLE K2
AISLE K3
AISLE K4
BONAVENTURE WAY
NOBLE JONES WAY
WILTBERGER WAY
MULLRYNE WAY
WALZ WAY
MYERS WAY
SHEFTALL WAY
LEGION WAY
EVERGREEN CIR.
PLANTATION SQUARE
NOBLE JONES DR.
MULLRYNE WAY
AISLE P1
AISLE P2
AISLE P3
AISLE P4
AISLE P5
AISLE P6
AISLE P7
AISLE P8
A
B
C
D
E
F
G
H
O
P
R
Richards
Comer
Butler
Dieter
Schwarz
Theus
Lubs
Wilson
McFarland
General Anderson
Myrick
Chisholm
Paine
Stobbs
Clinch
Zouberbuhler
Schroder
Lorch
Sneed
Wilbur
Estill Triangle
Estill
DeRenne
Backus
Schafer
Tiedman
Lester
Hohenstein
Ashmore
Arnold
Bryan
Foster
Stoddard
Herndon & H. Mercer
Houstoun
Little Gracie
Oldest burials
Roberts
Kollock
Taliaferro
Tattnall
Kollock
Hardee
Padelford
Herschbach
Noble Jones
Colonials at Bonaventure
Paulson
Colonial Monument Garden
Hodgson
Telfair
Lovell
Wiltberger
George Jones & Noble Wimberly Jones
American Legion Field
Herman
Guckenheimer
Meinhard
Jewish Circle
Jewish Chapel
Perlinski-Schaul
Frank
Rosenheim
Raskin
Myers

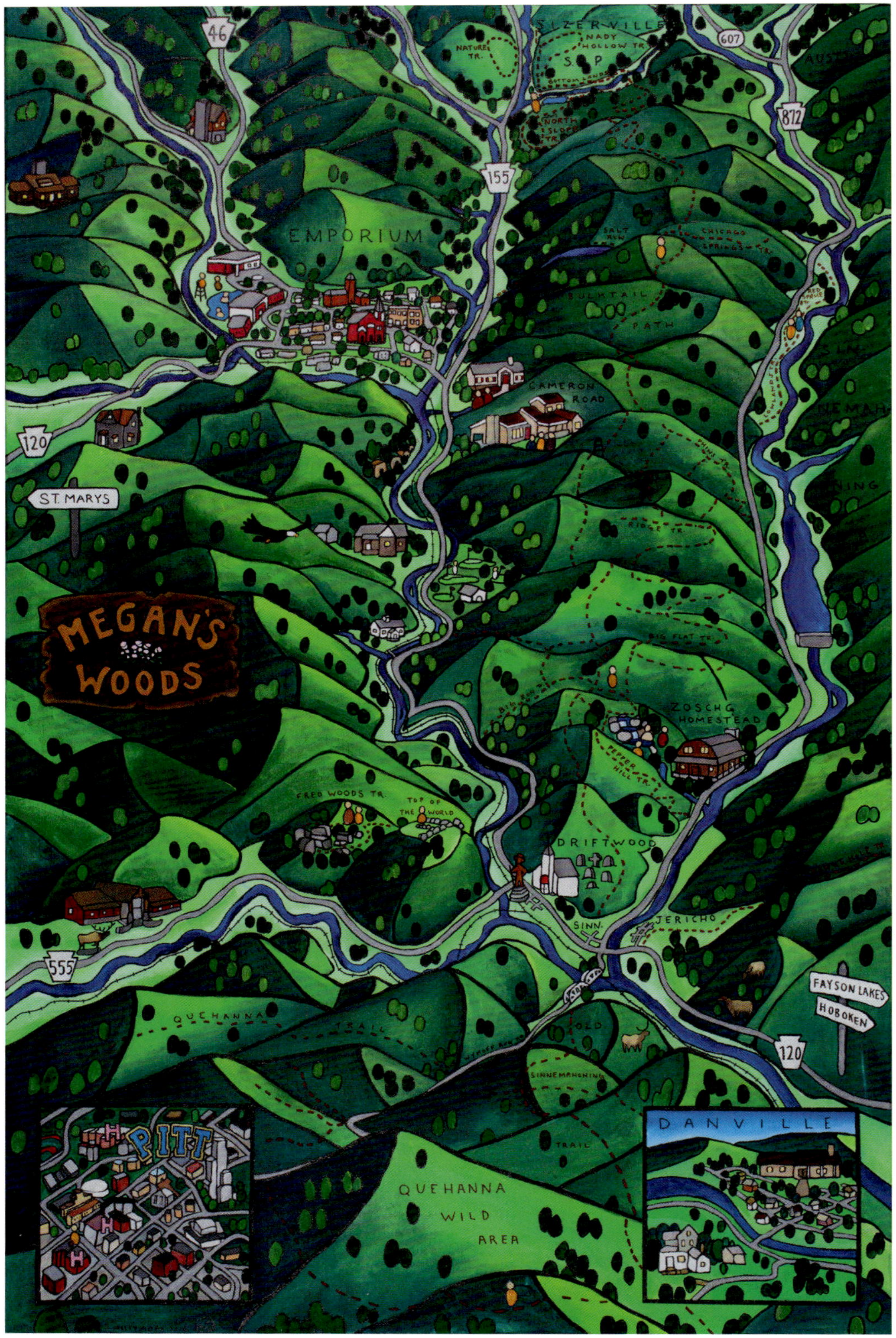

Megan's Woods »

Molly Holmberg Brown | MollyMaps

I remember thinking, multiple times, just tell him, "No." He wanted a giant custom map as a surprise gift for his wife. A giant map that would include every special place from her childhood, every trail she was fond of running, every relative's house in accurate detail. To make it worse, this was all in a part of the country I hardly knew. Clearly, this was that commission I always knew I'd eventually get—the kind I just wouldn't be able to pull off.

I finally agreed to make "Megan's Woods" if I could try a new style I'd been experimenting with, using acrylic markers and paints on canvas. The result would deviate from the qualities most maps are known for—subtlety, authoritativeness, fine detail, and soft color palettes. Instead, this map would embrace youthful simplicity, bright colors, and bold lines.

As whimsical as the final painting looks, the process of making it was highly analytical. I integrated topographic maps, lengthy client notes, trail descriptions, old photographs, digital imagery, and phone conversations to shape each turn in the road, every slope in the topography, each building color. Once all the information was gathered, I began composing it within the confines of a simple, rectangular shape. I had to shrink and expand geographical features, tinker with perspective, and adjust distances. Like all cartographers, I selected, simplified, omitted, and distorted in order to tell the best visual story I could.

I love the challenge of creating custom mental maps, crafting the best possible fusion of the experiential perception of place and Cartesian knowledge of the physical landscape. In all maps—but even more in mental maps—there is no clear distinction at the outset between what is important and what is not. You must sketch and study and chat and visualize again and again until you feel the urge, despite all the compositional constraints and distortions, to draw just the right bend in the river. Ultimately, my only judge is the person whose sense of place inspired the map. In this case, only Megan can tell me if it feels right, if I did my job.

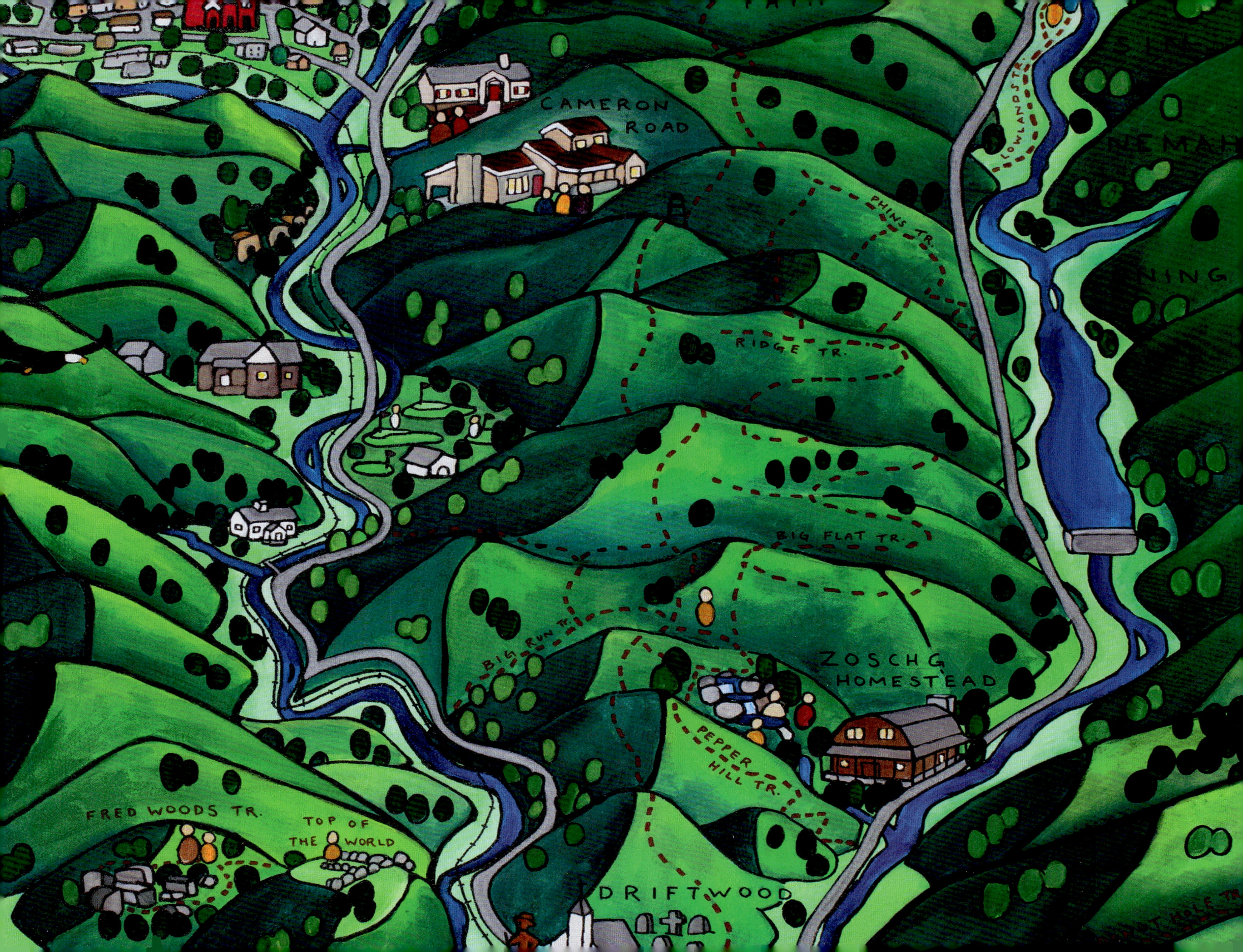
CAMERON ROAD
LOWLANDS TR.
PHINS TR.
RIDGE TR.
BIG FLAT TR.
BIG RUN TR.
ZOSCHG HOMESTEAD
PEPPER HILL TR.
FRED WOODS TR.
TOP OF THE WORLD
DRIFTWOOD

Traditionally, Martian craters are named after features on Earth, including small towns, scientists, and famous artists.

Here There Be Robots »

Eleanor Lutz

This map of Mars was inspired by maps made by medieval explorers. I love the hand-drawn detail of old maps and wanted to digitally recreate the beautiful calligraphy and ink work. My map isn't an exact replica of any particular past style, but the strongest influences are medieval and Victorian European maps.

In addition to the aesthetic qualities, I thought that an older style would be particularly fitting for a Mars map. Medieval and Victorian maps were made to describe faraway lands on Earth at a time when most of the world remained a mystery and most people only knew about the geography directly adjacent to their regions. Now that people have accurate maps to describe the entire globe in minute detail, our neighboring planets have taken over our imaginations as the next mysterious landscapes to explore. I wanted to use this thematic connection to tie together the history of Earth exploration with new adventures to extraterrestrial lands.

To make this map I drew the topographic layers of Mars by hand to give the map a traditional aesthetic. Then I added an overlay of actual NASA satellite imagery to provide additional texture. Traditionally, Martian craters are named after features on Earth, including small towns, scientists, and famous artists. In this map I tried to make a strange planet seem familiar by sharing the Earth origins of each crater name. I especially like the incongruity of Nigerian towns located next to French, Austrian, and Mexican towns.

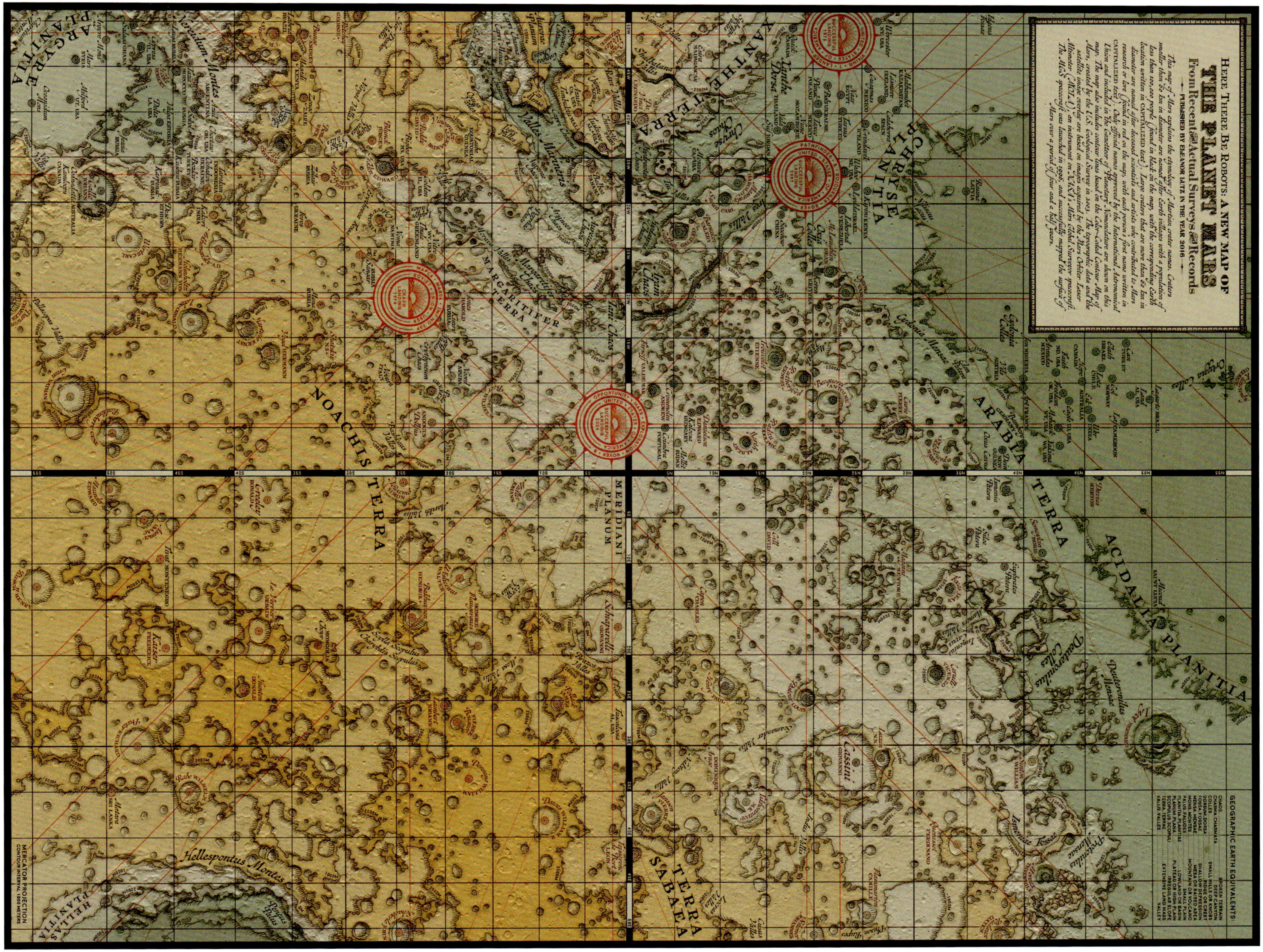

HERE THERE BE ROBOTS: A NEW MAP OF
THE PLANET MARS
From Recent & Actual Surveys & Records
PUBLISHED BY ELEANOR LUTZ IN THE YEAR 2016
This map of Mars explains the etymology of Martian crater names. Craters smaller than 60 km in diameter are named after Earth villages with a population of less than 100,000 people (found in black on the map, with the corresponding Earth location written in CAPITALIZED text). Large craters that are more than 60 km in diameter are named after deceased scientists and artists who contributed to Mars research or lore (found in red on the map, with each person's first name written in CAPITALIZED text). Only official names approved by the International Astronomical Union and included in the Gazetteer of Planetary Nomenclature are shown on this map. The map also includes contour lines based on the Color-Coded Contour Map of Mars, created by the U.S. Geological Survey in 2003. The topographic data and the satellite image overlay are based on images acquired by the Mars Orbiter Laser Altimeter (MOLA), an instrument on NASA's Mars Global Surveyor spacecraft. The MGS spacecraft was launched in 1996, and successfully mapped the surface of Mars over a period of four and a half years.
GEOGRAPHIC EARTH EQUIVALENTS:
CHAOS ... BROKEN TERRAIN
CHASMA, CHASMATA ... DEEP CANYON
COLLES ... SMALL HILLS OR KNOBS
DORSUM, DORSA ... RIDGE OR CREST
FOSSA, FOSSAE ... SHALLOW DEPRESSION
MENSA, MENSAE ... MESA OR RAISED AREA
MONS, MONTES ... MOUNTAIN OR VOLCANO
PALUS, PALUDES ... SMALL PLAIN
PLANITIA, PLANITIAE ... LOWLAND OR BASIN
PLANUM, PLANA ... PLATEAU OR HIGH PLAIN
SCOPULUS, SCOPULI ... IRREGULAR SLOPE
TERRA, TERRAE ... EXTENSIVE LAND MASS
VALLIS, VALLES ... VALLEY
MERCATOR PROJECTION
CONTOUR INTERVAL 1000 METERS
ARGYRE PLANITIA
XANTHE TERRA
CHRYSE PLANITIA
MARGARITIFER TERRA
NOACHIS TERRA
MERIDIANI PLANUM
ARABIA TERRA
ACIDALIA PLANITIA
TERRA SABAEA
HELLAS PLANITIA
Hellespontus Montes
Valles Marineris
Cassini
Schiaparelli
Deuteronilus Mensae

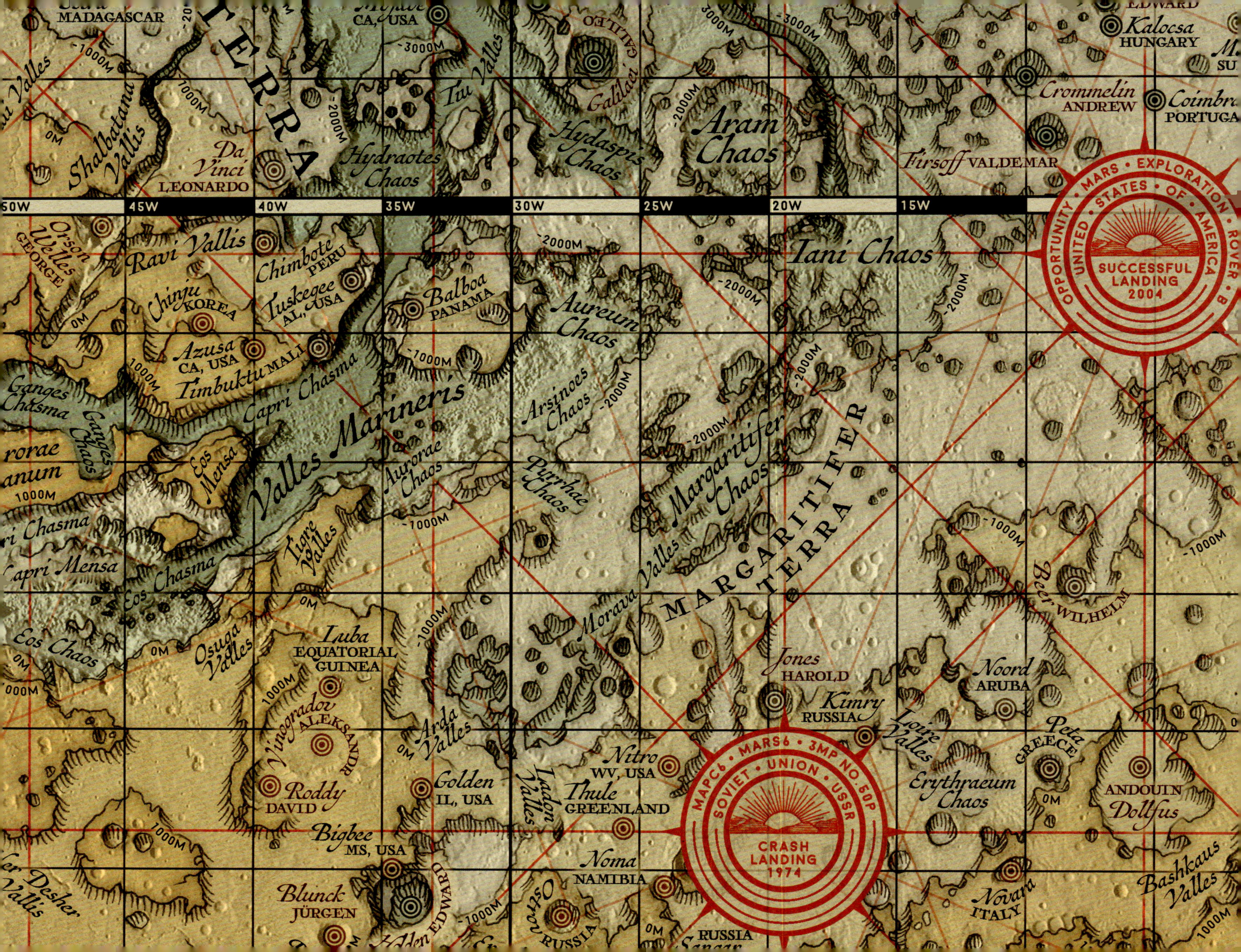
MADAGASCAR
CA, USA
TERRA
Galilaei
GALILEO
Tiu Valles
Hydraotes Chaos
Hydaspis Chaos
Aram Chaos
Kalocsa
HUNGARY
EDWARD
Crommelin
ANDREW
Coimbra
PORTUGAL
Firsoff
VALDEMAR
Shalbatana Vallis
Da Vinci
LEONARDO
50W
45W
40W
35W
30W
25W
20W
15W
Orson Welles
GEORGE
Ravi Vallis
Chimbote
PERU
Tuskegee
AL, USA
Balboa
PANAMA
Chinju
KOREA
Azusa
CA, USA
Timbuktu
MALI
Capri Chasma
Aureum Chaos
Iani Chaos
Arsinoes Chaos
Ganges Chasma
Ganges Chaos
Eos Mensa
Valles Marineris
Aurorae Chaos
Pyrrhae Chaos
Margaritifer Chaos
MARGARITIFER TERRA
Capri Mensa
Eos Chasma
Tiare Valles
Morava Valles
Eos Chaos
Osuga Valles
Luba
EQUATORIAL GUINEA
Jones
HAROLD
Kimry
RUSSIA
Noord
ARUBA
Beer
WILHELM
Vinogradov
ALEKSANDR
Arda Valles
Nitro
WV, USA
Loire Valles
Peta
GREECE
Roddy
DAVID
Golden
IL, USA
Ladon Valles
Thule
GREENLAND
Erythraeum Chaos
ANDOUIN
Dollfus
Bigbee
MS, USA
Noma
NAMIBIA
Blunck
JÜRGEN
Holden
EDWARD
Ostrov
RUSSIA
Novara
ITALY
Bashkaus Valles
Desher Vallis
UNITED STATES OF AMERICA
MARS EXPLORATION ROVER
OPPORTUNITY
SUCCESSFUL LANDING 2004
SOVIET UNION USSR
MAPC6 3MP NO 50P
CRASH LANDING 1974
-1000M
-2000M
-3000M
0M
1000M

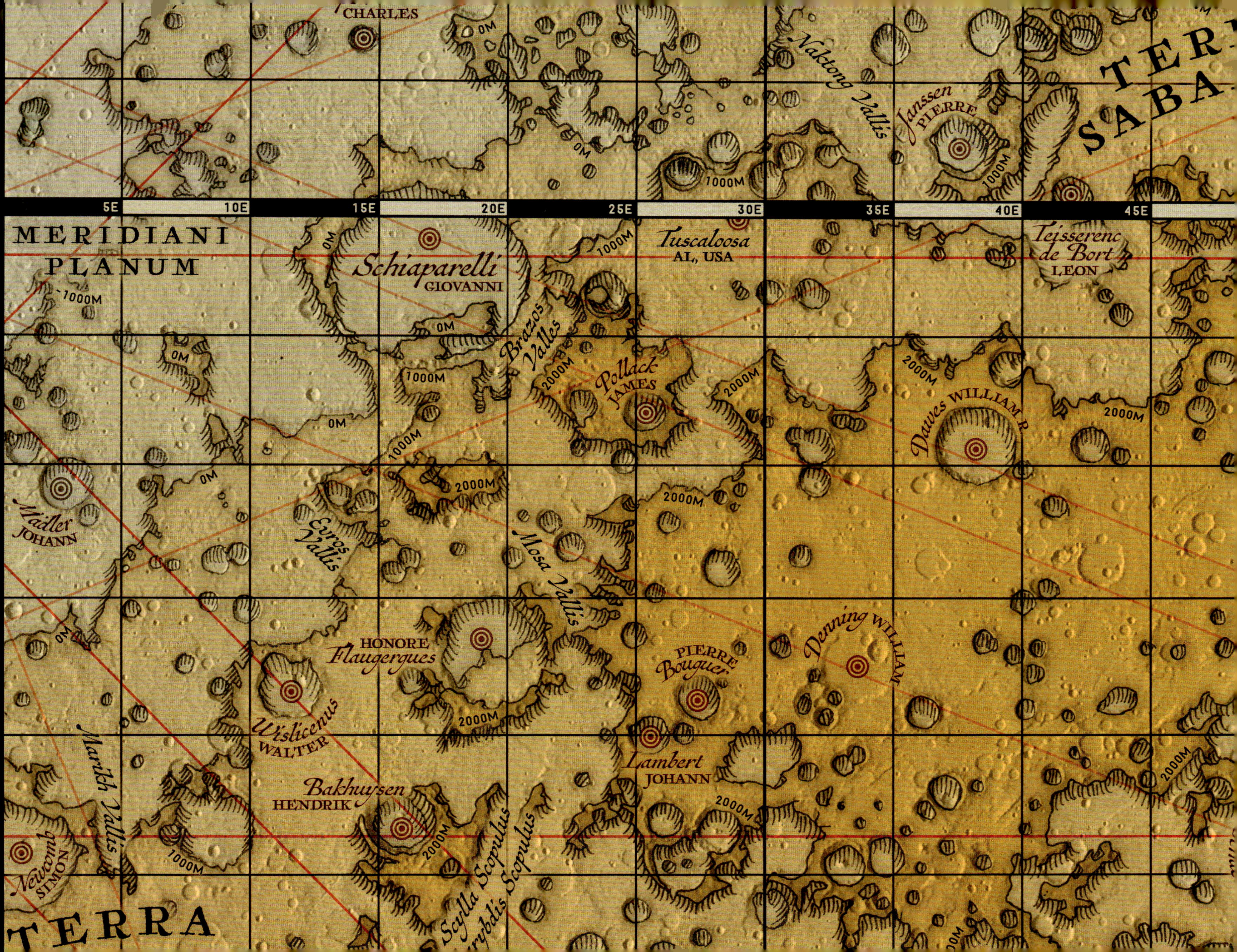
CHARLES
Naktong Vallis
TER
SABA
Janssen
PIERRE
5E
10E
15E
20E
25E
30E
35E
40E
45E
MERIDIANI
PLANUM
Schiaparelli
GIOVANNI
Tuscaloosa
AL, USA
Teisserenc
de Bort
LEON
Brazos Valles
Pollack
JAMES
Dawes WILLIAM R
Mädler
JOHANN
Evros Vallis
Mosa Vallis
HONORE
Flaugergues
PIERRE
Bouguer
Denning WILLIAM
Wislicenus
WALTER
Lambert
JOHANN
Bakhuysen
HENDRIK
Marikh Vallis
Newcomb
SIMON
Scylla Scopulus
rybdis Scopulus
TERRA
0M
1000M
2000M

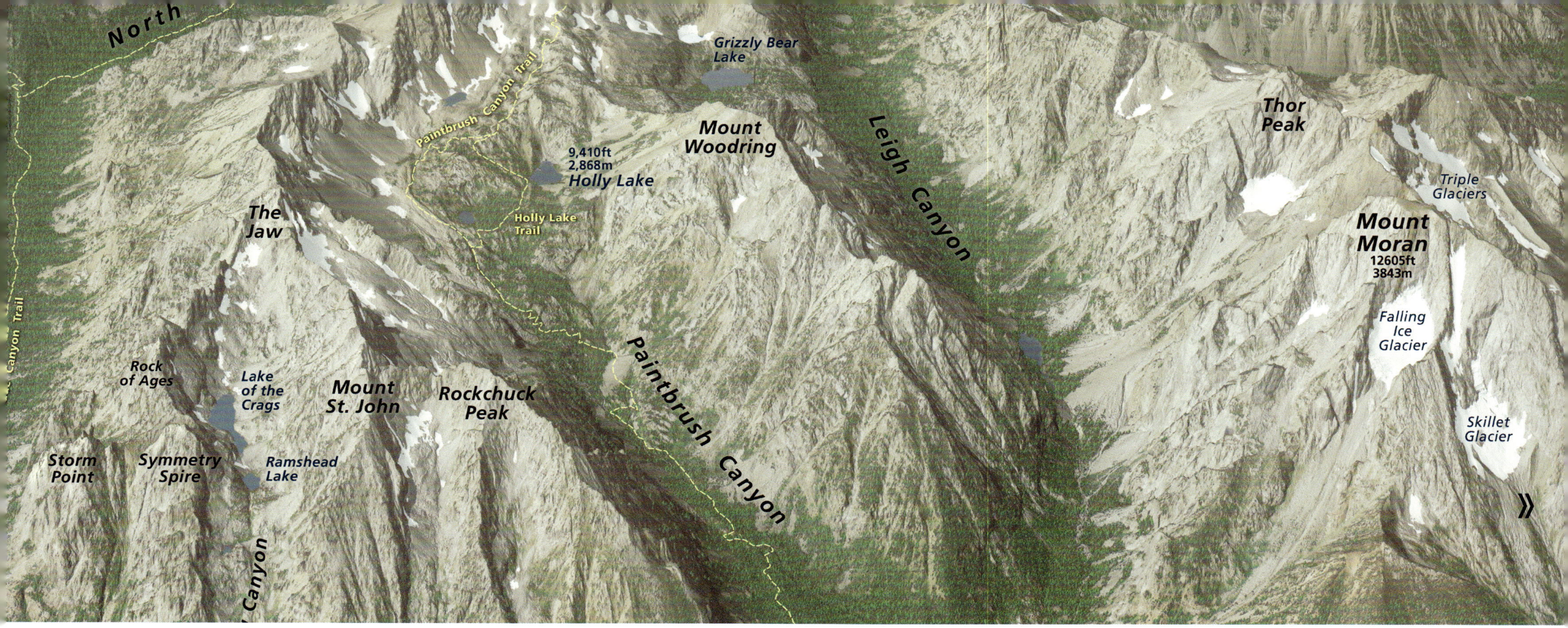

Jenny Lake Hiking Trails »

Tom Patterson

Designing a successful 3-D oblique terrain map depends on strict conditions. You must have a valid reason, an appropriate audience, and proper data for making a 3-D map. The terrain itself must be conducive to three-dimensional depiction, and you need ample time to produce the 3-D map, which is considerably more involved than producing a map in 2-D shaded relief. In the case of my map of the Jenny Lake Hiking Trails in Grand Teton National Park, I got lucky on all of these counts.

I made this map for the Jenny Lake Visitor Center. It is on display at an outdoor kiosk facing west toward the imposing Teton Range. To reveal the mountain trails, I selected a 3-D camera position beyond the Visitor Center, looking west at high elevation and looking down on the Tetons at a relatively steep angle. One problem with this camera position was the absence of silhouetted mountains against the sky, a trait of classic alpine panoramas. To compensate, I added contrast to the summits to make them pop. I also applied generous background haze to enhance the sense of depth.

So that visitors could better relate the map to the landscape, I rendered the map with realistic colors and textures, but with modifications to improve legibility. For example, the forest textures are a much lighter green than they appear naturally. I also lightened the dark cast shadows that obscured detail on the steepest slopes, and allowed snow patches and glaciers to glisten more. I paid close attention to map labels so that they are large and contrast with the background terrain art enough for comfortable reading from several feet away. I applied barely perceptible blurring and lightening to the terrain art underneath the labels for better legibility. Finally, I painted a sun glint on Jenny Lake to direct viewers to their location—and because it looked so nice.

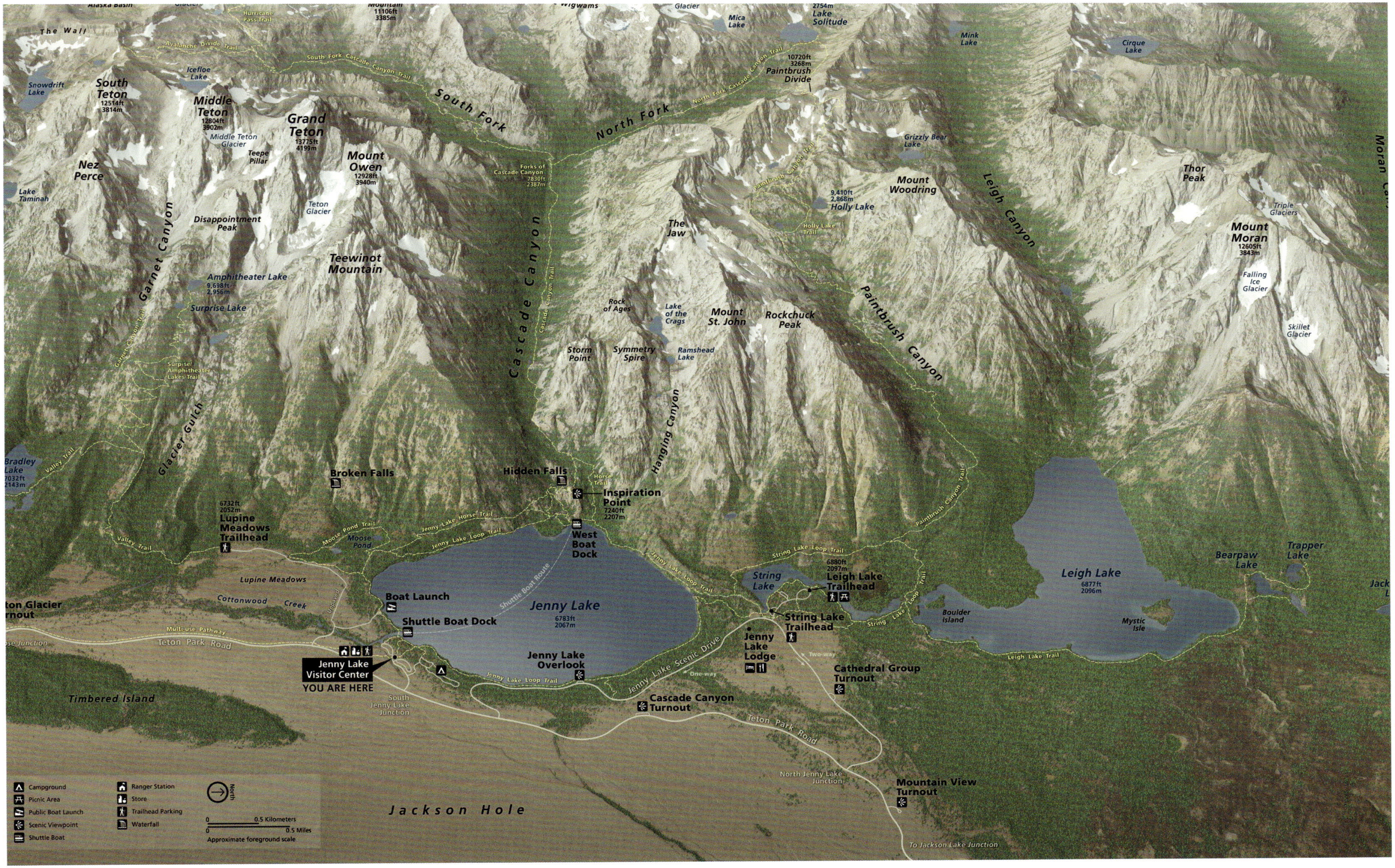

Alaska Basin
The Wall
Hurricane Pass Trail
Avalanche Divide Trail
South Fork Cascade Canyon Trail
Snowdrift Lake
South Teton
12514ft
3814m
Icefloe Lake
Middle Teton
12804ft
3902m
Middle Teton Glacier
Grand Teton
13775ft
4199m
Teepe Pillar
Nez Perce
Lake Taminah
Mount Owen
12928ft
3940m
Teton Glacier
Disappointment Peak
Garnet Canyon
Teewinot Mountain
Amphitheater Lake
9,698ft
2,956m
Surprise Lake
Surprise/Amphitheater Lakes Trail
Garnet Canyon Trail
Glacier Gulch
Valley Trail
Bradley Lake
7032ft
2143m
South Fork
North Fork
Forks of Cascade Canyon
7830ft
2387m
Cascade Canyon
Cascade Canyon Trail
Wigwams
Mica Lake
Lake Solitude
10720ft
3268m
Paintbrush Divide
North Fork Cascade Canyon Trail
Mink Lake
Cirque Lake
Grizzly Bear Lake
Mount Woodring
9,410ft
2,868m
Holly Lake
Holly Lake Trail
Paintbrush Canyon Trail
Leigh Canyon
Thor Peak
Moran Canyon
Triple Glaciers
Mount Moran
12605ft
3843m
Falling Ice Glacier
Skillet Glacier
The Jaw
Rock of Ages
Lake of the Crags
Mount St. John
Rockchuck Peak
Paintbrush Canyon
Storm Point
Symmetry Spire
Ramshead Lake
Hanging Canyon
Broken Falls
Hidden Falls
Horse Trail
Inspiration Point
7240ft
2207m
6732ft
2052m
Lupine Meadows Trailhead
Moose Pond Trail
Moose Pond
Jenny Lake Horse Trail
Jenny Lake Loop Trail
West Boat Dock
Shuttle Boat Route
Jenny Lake
6783ft
2067m
String Lake Loop Trail
String Lake
6880ft
2097m
Leigh Lake Trailhead
String Lake Trailhead
Leigh Lake
6877ft
2096m
Boulder Island
Mystic Isle
Leigh Lake Trail
Bearpaw Lake
Trapper Lake
Lupine Meadows
Cottonwood Creek
Glacier Gulch Turnout
Multi-use Pathway
Teton Park Road
Moose Junction
Boat Launch
Shuttle Boat Dock
Jenny Lake Visitor Center
YOU ARE HERE
South Jenny Lake Junction
Timbered Island
Jenny Lake Overlook
Jenny Lake Scenic Drive
One-way
Two-way
Jenny Lake Lodge
Cathedral Group Turnout
Cascade Canyon Turnout
North Jenny Lake Junction
Mountain View Turnout
To Jackson Lake Junction
Jackson Hole
Campground
Picnic Area
Public Boat Launch
Scenic Viewpoint
Shuttle Boat
Ranger Station
Store
Trailhead Parking
Waterfall
North
0
0.5 Kilometers
0
0.5 Miles
Approximate foreground scale

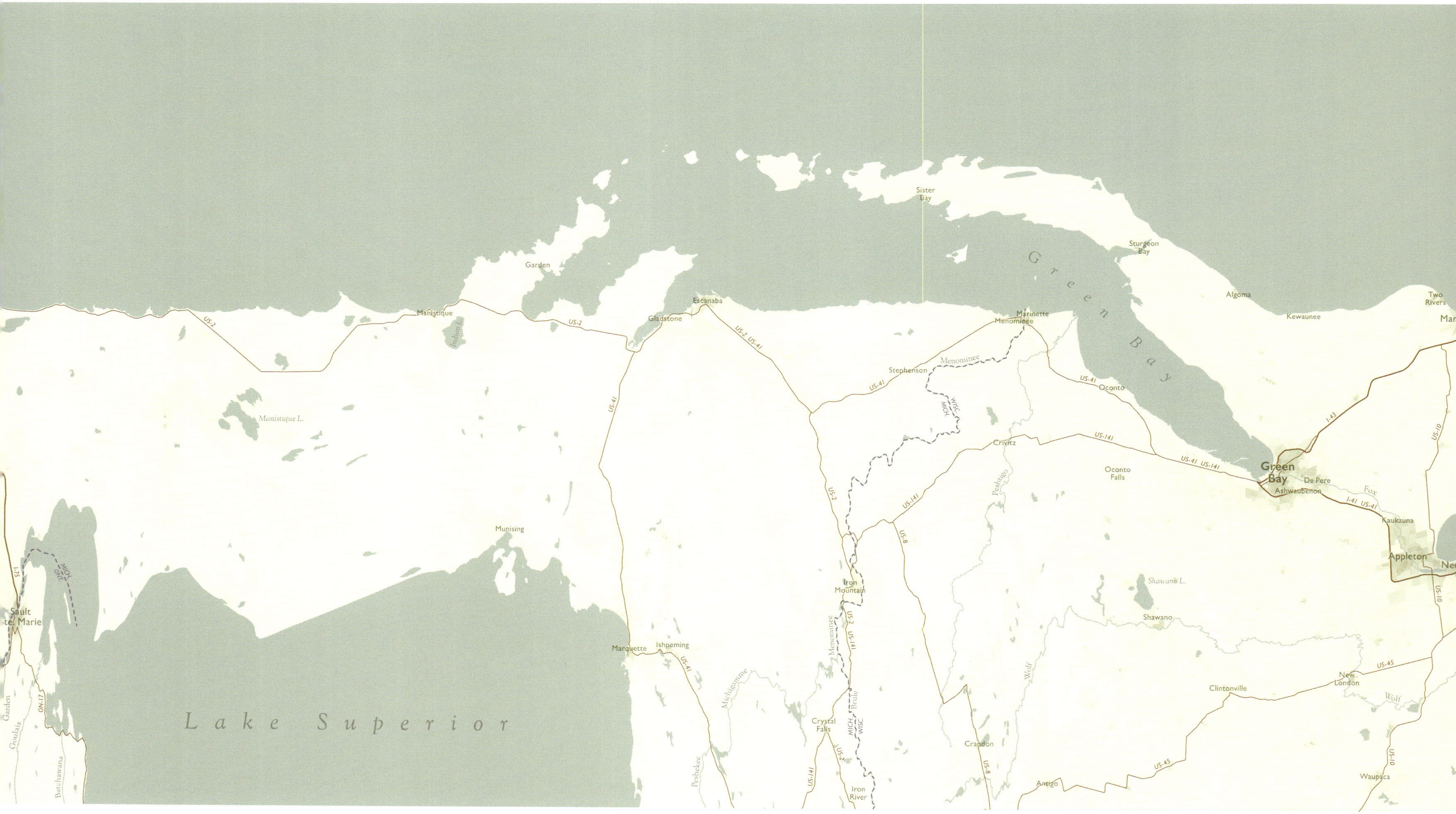
Sister Bay
Sturgeon Bay
Green Bay
Garden
Algoma
Two Rivers
Manistique
Escanaba
Gladstone
Marinette
Menominee
Kewaunee
Manit
US-2
Indian L.
US-2 US-41
Stephenson
Menominee
US-41
Oconto
WISC. MICH.
Manistique L.
US-41
I-43
US-10
US-141
Crivitz
US-41 US-141
Oconto Falls
Green Bay
De Pere
Ashwaubenon
Fox
I-41 US-41
Kaukauna
US-2
US-141
Peshtigo
US-8
Munising
Appleton
Neen
MICH. ONT.
Iron Mountain
Shawano L.
Sault Ste. Marie
US-2 US-141
Shawano
Marquette
Ishpeming
US-41
Menominee
Wolf
US-45
New London
Clintonville
Wolf
Lake Superior
Michigamme
Crystal Falls
Brule
MICH. WISC.
Crandon
Garden
Goulais
ON-17
Batchawana
Peshekee
US-141
US-2
US-8
US-45
US-10
Iron River
Antigo
Waupaca
I-75

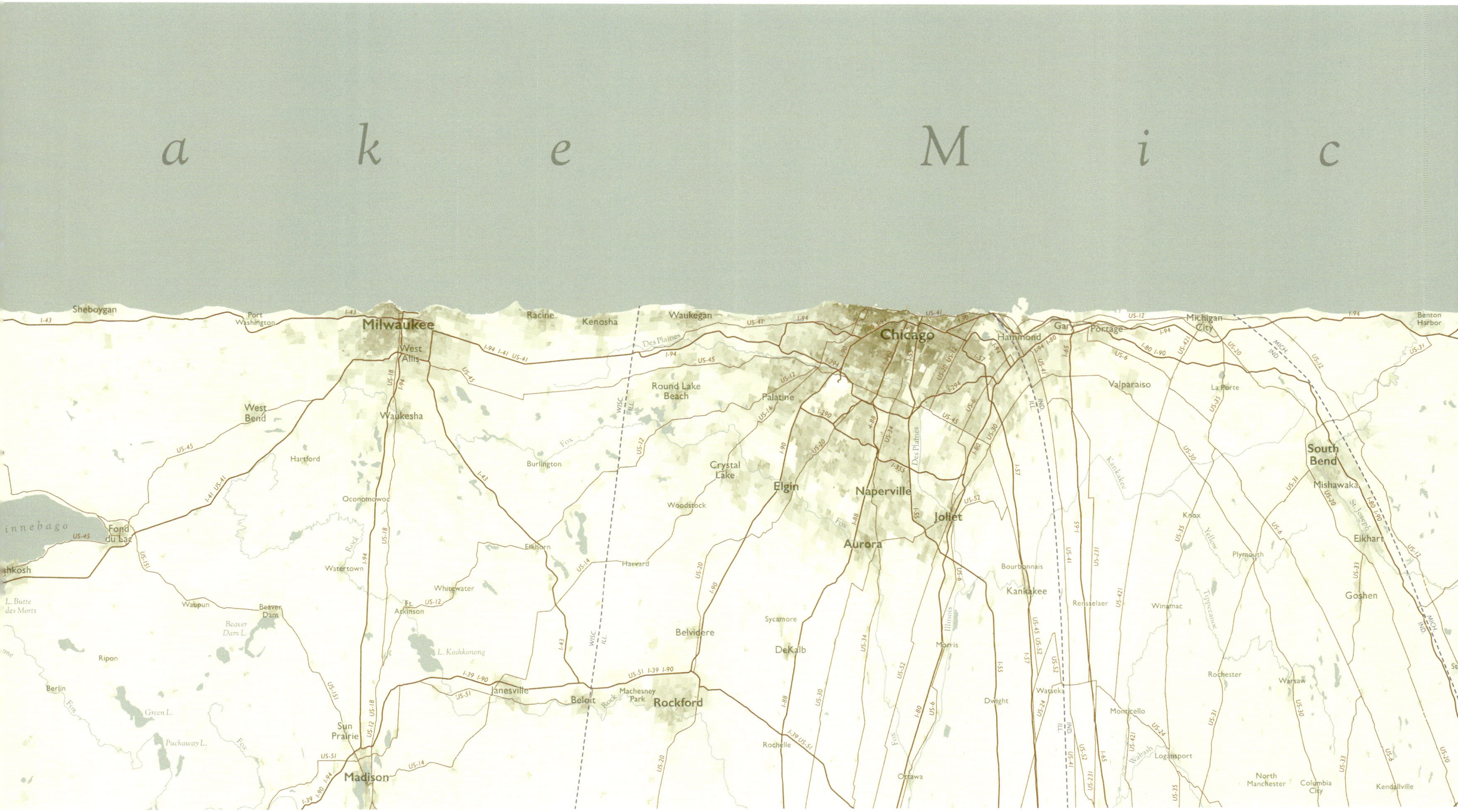
a k e M i c
Sheboygan
Port Washington
Milwaukee
West Allis
Racine
Kenosha
Waukegan
Chicago
Hammond
Gary
Portage
Michigan City
Benton Harbor
Round Lake Beach
Palatine
Waukesha
West Bend
Hartford
Burlington
Crystal Lake
Elgin
Naperville
Joliet
Aurora
Woodstock
Valparaiso
La Porte
South Bend
Mishawaka
Elkhart
Goshen
Plymouth
Knox
Oconomowoc
Fond du Lac
Oshkosh
Watertown
Elkhorn
Harvard
Whitewater
Ft. Atkinson
Waupun
Beaver Dam
Beaver Dam L.
L. Koshkonong
Ripon
Berlin
Green L.
Puckaway L.
Sycamore
DeKalb
Belvidere
Janesville
Beloit
Machesney Park
Rockford
Sun Prairie
Madison
Rochelle
Ottawa
Morris
Dwight
Bourbonnais
Kankakee
Rensselaer
Winamac
Watseka
Monticello
Logansport
Rochester
Warsaw
North Manchester
Columbia City
Kendallville
WISC ILL
MICH IND
IND ILL

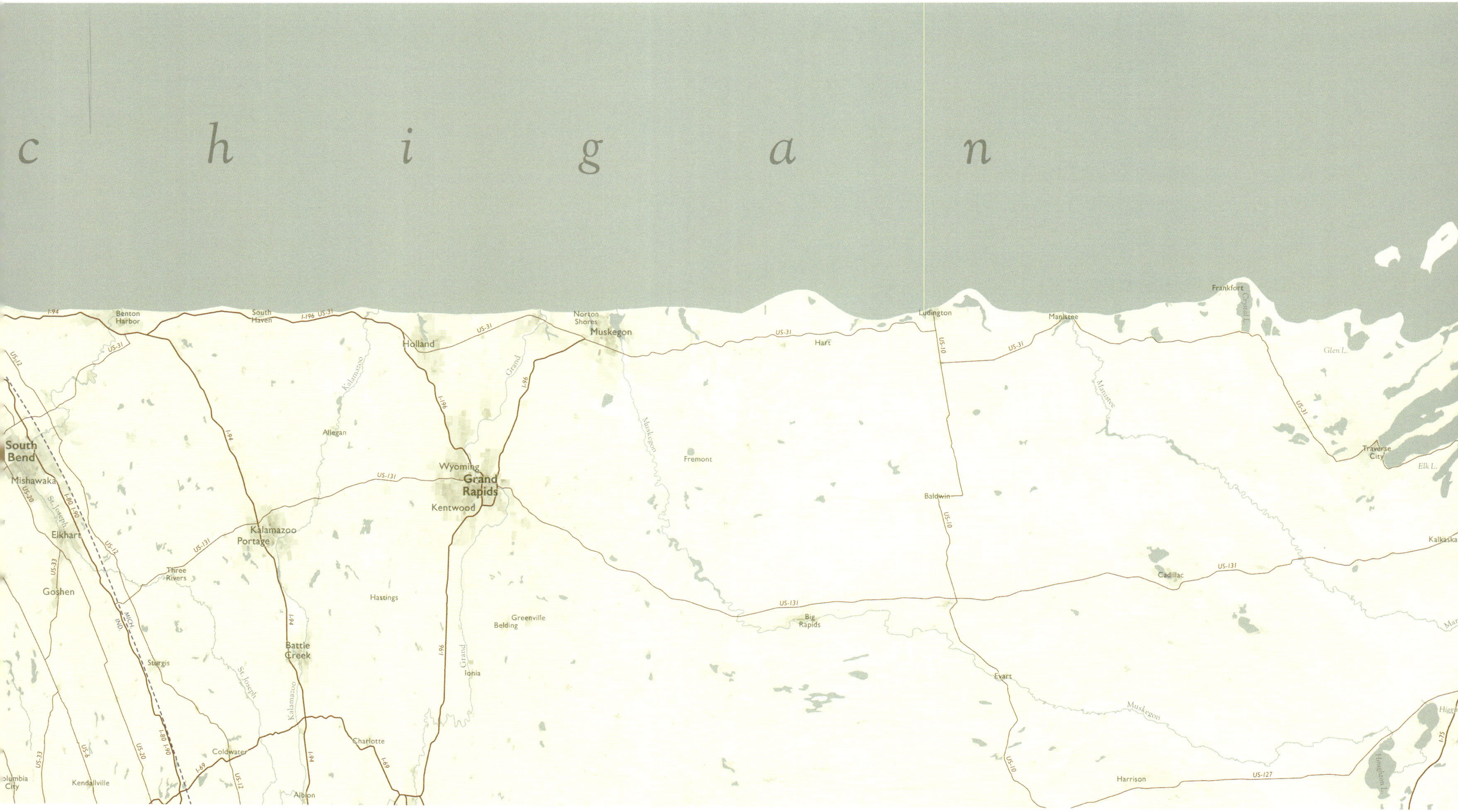
c h i g a n
Benton Harbor
South Haven
Norton Shores
Muskegon
Holland
Ludington
Hart
Manistee
Frankfort
Glen L.
Traverse City
Elk L.
South Bend
Mishawaka
Elkhart
Goshen
Allegan
Wyoming
Grand Rapids
Kentwood
Kalamazoo
Portage
Three Rivers
Hastings
Greenville
Belding
Battle Creek
Sturgis
Ionia
Charlotte
Coldwater
Kendallville
Albion
Fremont
Baldwin
Big Rapids
Cadillac
Kalkaska
Evart
Harrison
Higgins
US-31
US-10
US-131
US-127
US-12
US-20
US-33
US-6
I-94
I-196
I-96
I-69
I-75
I-80 I-90
MICH
IND
Grand
Kalamazoo
Muskegon
Manistee
St. Joseph

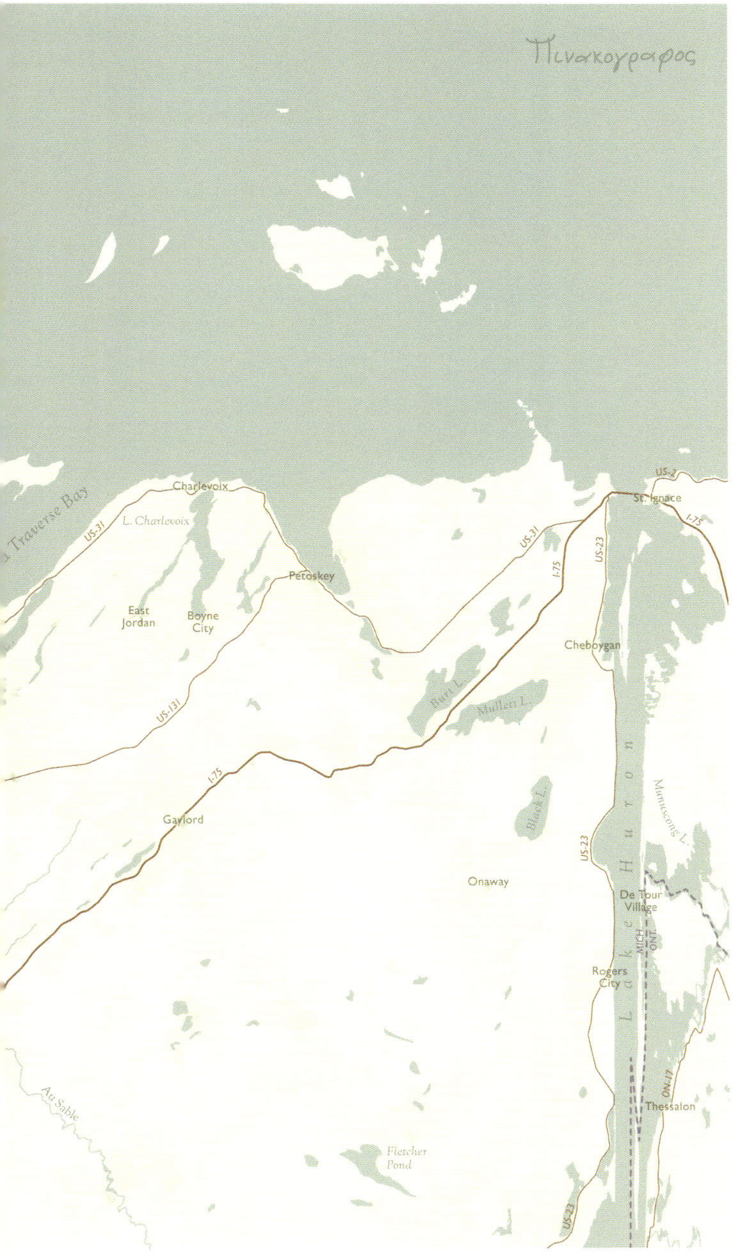

« An Unfurling of Lake Michigan

Daniel P. Huffman

One evening, a conversation with a colleague led me to wonder: What would happen if you took the shoreline of a lake, cut it, and straightened it out? Freed from the tyranny of latitude and longitude, we can instead arrange geography according to a sort of lake-based coordinate system, with axes that measure the distance from, or along, the shore.

The resulting map offers an unusual perspective, but one that also feels deeply familiar: I have lived most of my life within this strip of land along the Lake Michigan shore. I orient myself according to the inland sea, the anchor of my personal geography. This map depicts an experiential view of the region—one which, I suspect, resonates with many Great Lakes denizens. It is a human perspective that conveys just how far away my former home in Michigan feels from my current one in Wisconsin, when the route between them involves traveling around the lake through Illinois.

So while it is unusual to view the world this way, it is perfectly appropriate to do so, which is something I wanted to capture aesthetically. My goal was to "naturalize" this perspective, by giving the map a classic and understated look. It would have been easy to go in the opposite direction: playing up the weird distortions by pairing them with catchy typography and bold colors. But I did not want this perspective to be seen as weird. It's unusual, to be sure, but not abnormal.

The warm colors are meant to feel familiar and organic, like something you'd see in any run-of-the-mill map. Likewise, the typography: I used Gill Sans for the city and highway labels. It's a classic, lending an attractive and understated normality by rooting the map in the mid-twentieth century. I paired it with Sorts Mill Goudy for the water labels.

Cartographers often work to minimize distortions, but this project demonstrates one of the many ways in which those distortions can be leveraged to convey a valuable perspective: Humans do not always experience the world in latitude and longitude.

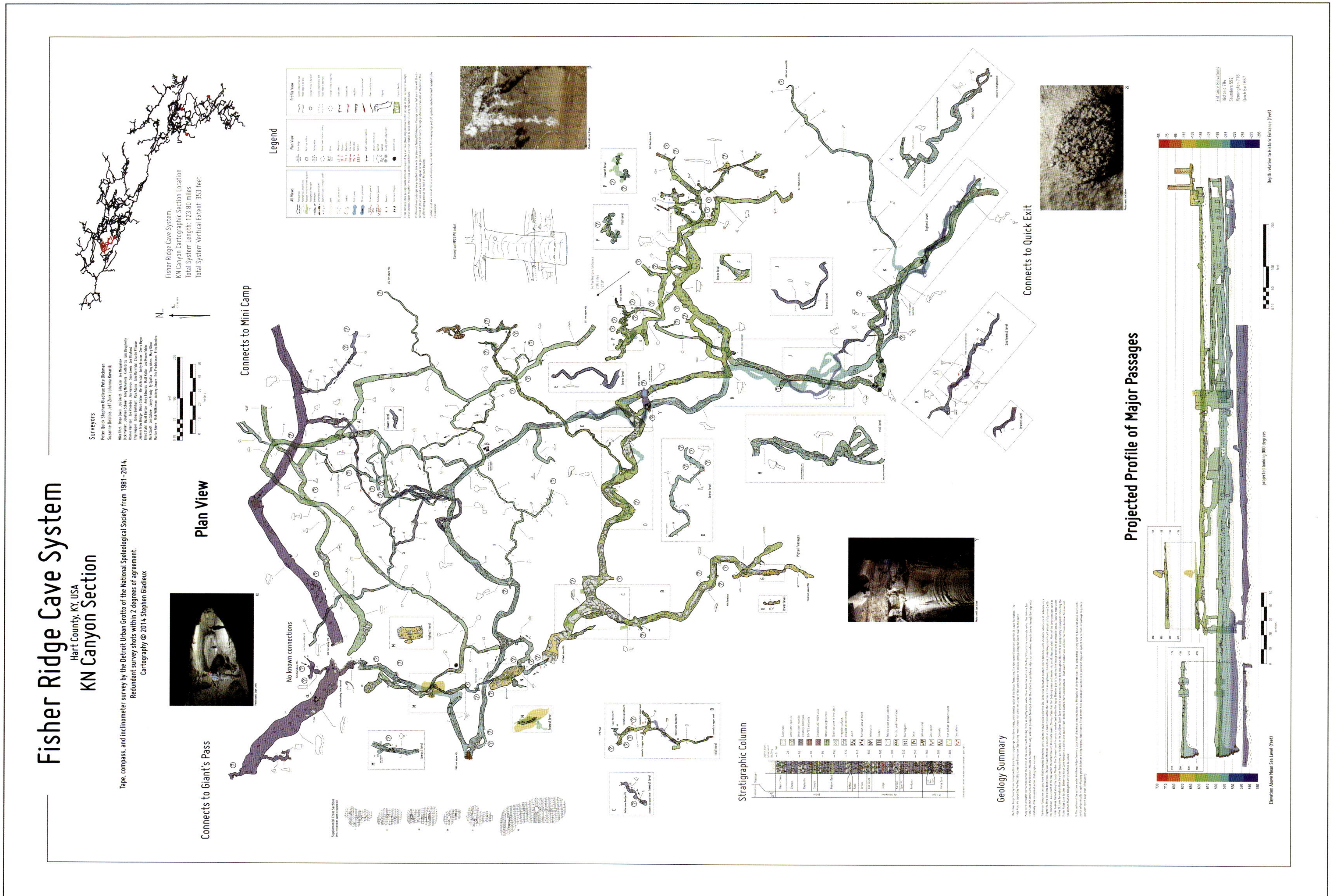
Fisher Ridge Cave System
Hart County, KY, USA
KN Canyon Section
Tape, compass, and inclinometer survey by the Detroit Urban Grotto of the National Speleological Society from 1981-2014.
Redundant survey shots within 2 degrees of agreement.
Cartography © 2014 Stephen Gladieux
Surveyors
Peter Quick Stephen Gladieux Pete Dickman
Suzanne Deblois Jeff Zink Johanna Kovarik
Fisher Ridge Cave System,
KN Canyon Cartographic Section Location
Total System Length: 123.80 miles
Total System Vertical Extent: 353 feet
Connects to Giant's Pass
Plan View
Connects to Mini Camp
No known connections
Legend
Stratigraphic Column
Geology Summary
Connects to Quick Exit
Projected Profile of Major Passages
Entrance Elevations
Historic 784
Saunders 592
Remington 716
Quick Exit 667
Elevation Above Mean Sea Level (feet)
projected looking 000 degrees
Depth relative to Historic Entrance (feet)

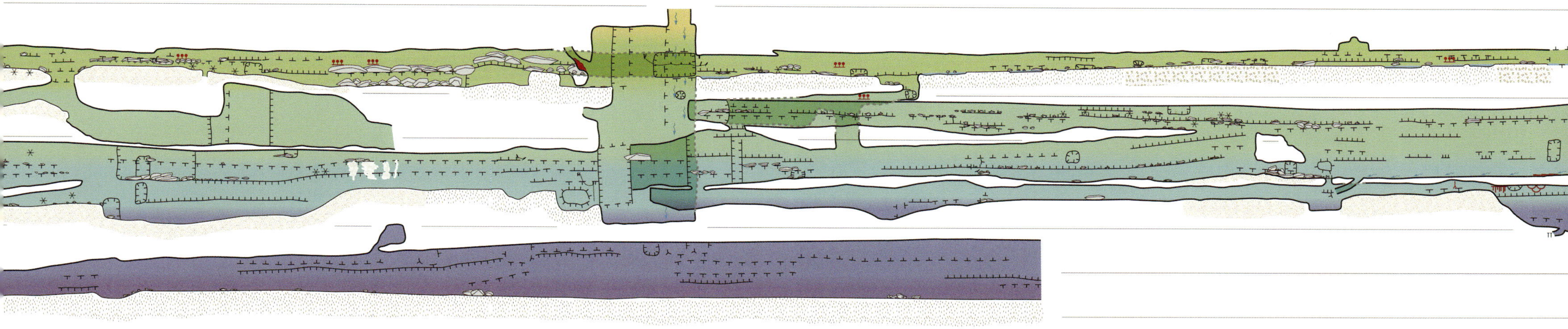

« Fisher Ridge Cave System, KN Canyon

Stephen Gladeux

Long before I began drawing maps I was fascinated by the vertical world. This composite cave map shows a vertically complex portion of Fisher Ridge Cave System in Kentucky. The map demonstrates that adding a third dimension presents unique challenges to a cartographer.

My goal was to create a tool that could be used in further exploration of this section of the cave system, and that required a map that displayed travel routes clearly enough to facilitate navigation for anyone unfamiliar with the area. I wanted the shape of the passage network to be most prominent. The known leads had to be displayed, and the layout in plan and profile had to hint at possible interconnectivity between known passages, exposing leads likely to result in significant new discoveries.

I used a projected profile to identify the main levels of development and then tied those levels into the plan with a rainbow spectrum, providing viewers with a sense of passage elevation. The profile delivers information about features on the near and far walls. My method combines fading colors that show gradual floor elevation change and sharp demarcations to indicate cliffs and other dramatic features. Directionally oriented cross sections give a sense of the difficulty of travel and local complexity. Photographs supplement the presentation to provide a more visceral sense of the cave environment. Plan cut-outs show overlying passage detail. I've also included conceptual, off-axis drawings to give clearer details related to rigging and travel.

Air flow in caves is a major indicator of passage development and lead potential. In this region of Kentucky, airflow is mainly controlled by the "chimney effect," or air density differences caused by temperature contrast between the surface and the cave. To provide additional useful information, I converted all of the raw flow notations to either "summer" or "winter" conditions based on whether the surface was warmer or colder on a particular date and time.

This map is now on display at the expedition headquarters and is used as an aid in exploration, constantly evolving as new passages are discovered.

Stratigraphic Column

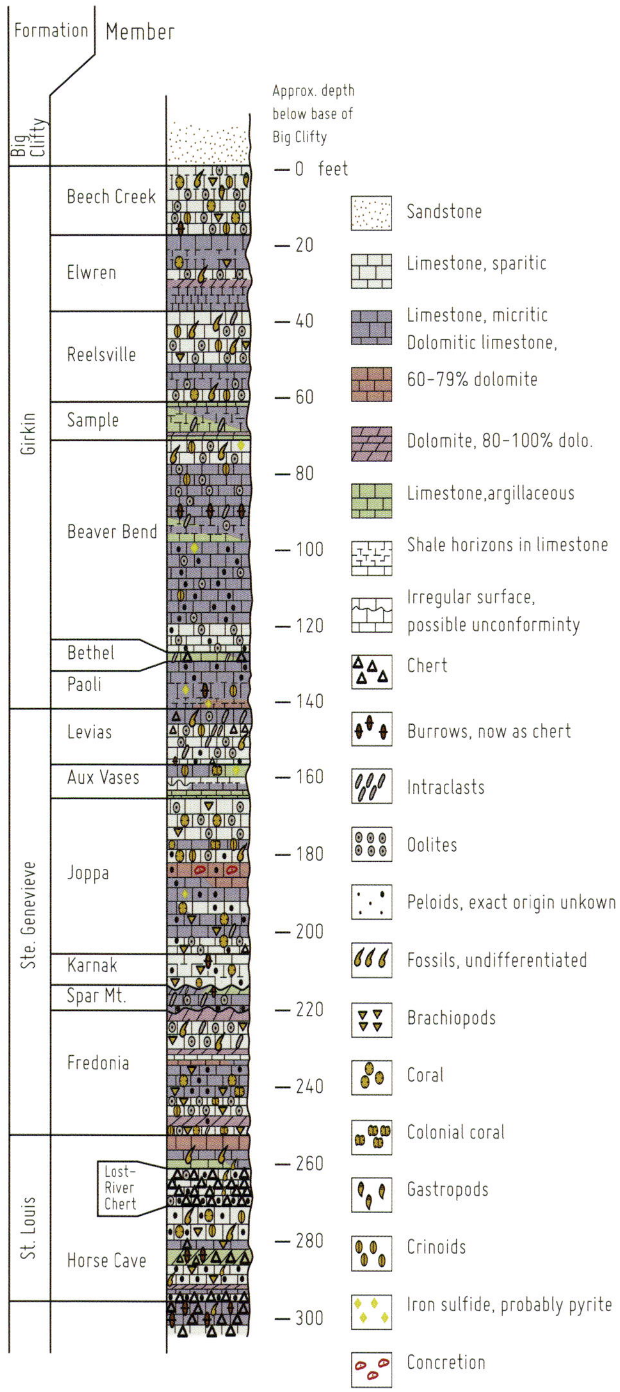

Stratagraphic column redrawn from Lawrence E. Bean 1987

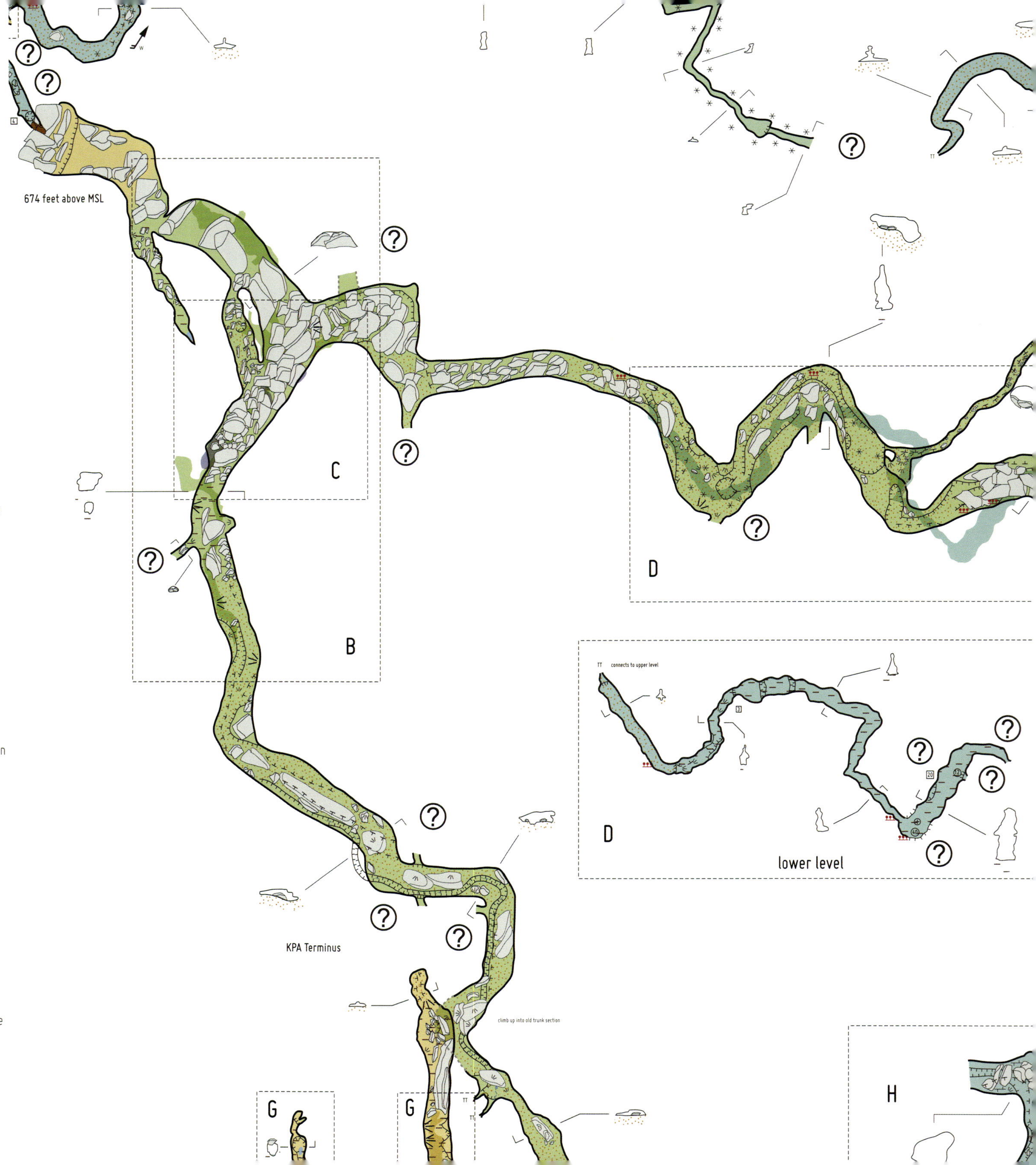

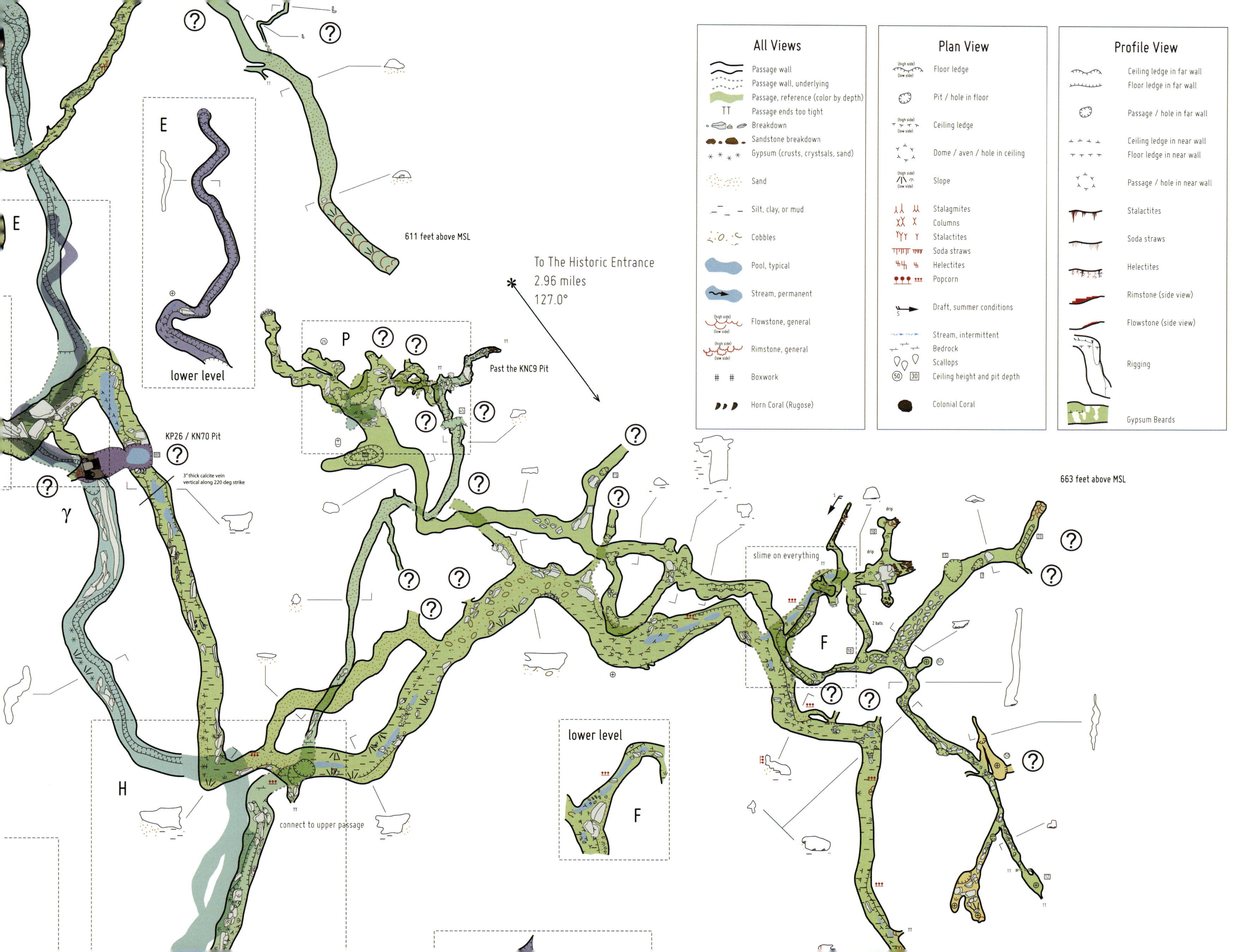

All Views
Passage wall
Passage wall, underlying
Passage, reference (color by depth)
Passage ends too tight
Breakdown
Sandstone breakdown
Gypsum (crusts, crystsals, sand)
Sand
Silt, clay, or mud
Cobbles
Pool, typical
Stream, permanent
Flowstone, general
Rimstone, general
Boxwork
Horn Coral (Rugose)
Plan View
Floor ledge
Pit / hole in floor
Ceiling ledge
Dome / aven / hole in ceiling
Slope
Stalagmites
Columns
Stalactites
Soda straws
Helectites
Popcorn
Draft, summer conditions
Stream, intermittent
Bedrock
Scallops
Ceiling height and pit depth
Colonial Coral
Profile View
Ceiling ledge in far wall
Floor ledge in far wall
Passage / hole in far wall
Ceiling ledge in near wall
Floor ledge in near wall
Passage / hole in near wall
Stalactites
Soda straws
Helectites
Rimstone (side view)
Flowstone (side view)
Rigging
Gypsum Beards
611 feet above MSL
To The Historic Entrance
2.96 miles
127.0°
E
lower level
E
P
Past the KNC9 Pit
KP26 / KN70 Pit
3" thick calcite vein
vertical along 220 deg strike
γ
663 feet above MSL
slime on everything
drip
2 bats
F
lower level
F
H
connect to upper passage

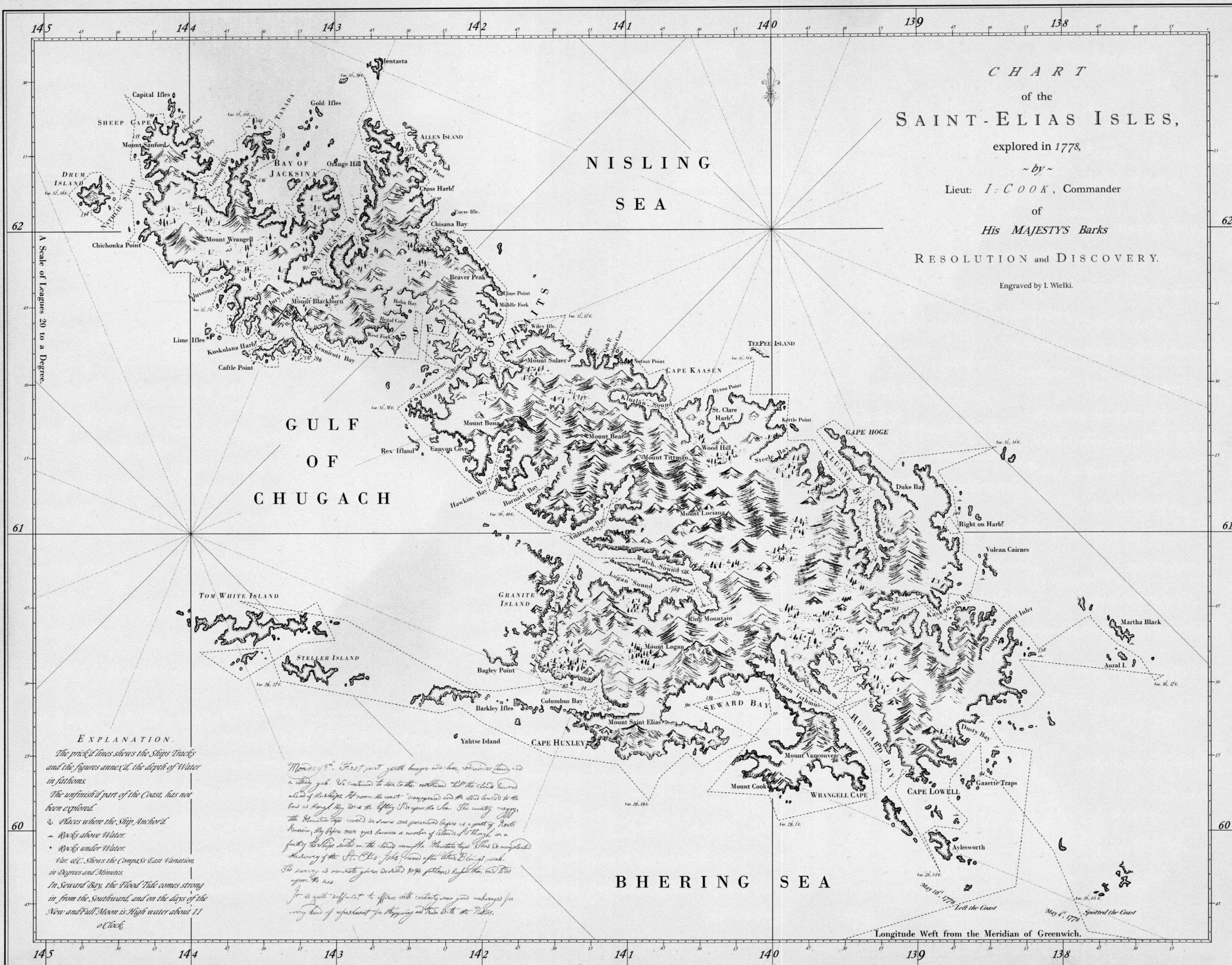

Published as the Act directs 1st Jan.y 1782.

《 Chart of the Saint-Elias Isles

Jeff Wielki

Occasionally, while hiking or skiing in the alpine, a valley cloud below can give the illusion that the mountain summits around you are islands in an ocean of clouds. What if an explorer could sail on that cloud ocean and map new lands? This was the initial inspiration for my map.

If James Cook had happened upon the St. Elias Mountain Range in Alaska and the Yukon on his 1778 voyage to chart the Northwest Passage and could have sailed on the clouds 2,000 meters above the Earth, he could have mapped these new islands in the same style as his stunning chart of New Zealand. Cook's New Zealand map is from an age in which monsters have been banished from mapmaking. The unknown remains unknown while scientific facts are gathered. Cook's map displays empty gaps in the coastline, contrasted with detailed harbours.

Cook's work reflects the scientific side of cartography, while still leaving it up to the viewer to discover the land. As a cartographer I wanted to emulate his map, and I wanted to discover something new.

The techniques used in compiling this map combine analog and digital, science and fantasy, reality and fiction. I sketched with pencil, by hand, using Digital Elevation Models and topographic maps as reference materials. I then scanned and digitally annotated the drawn shorelines and stylized mountains and terrain. That means the depth soundings, mountains, topography, and magnetic variation are all accurate for the date and location. Naturally, I used some artistic license too, given the style of the map. Gaps in the shoreline are based on distance and line of sight to reflect survey methods during Cook's era. Names are modern, but a valley or glacier may have become a bay or a strait.

The historical, fictional cartography of the map extends my inspirational concept far beyond ideas conceived while out of breath and lightheaded on a mountain ridge.

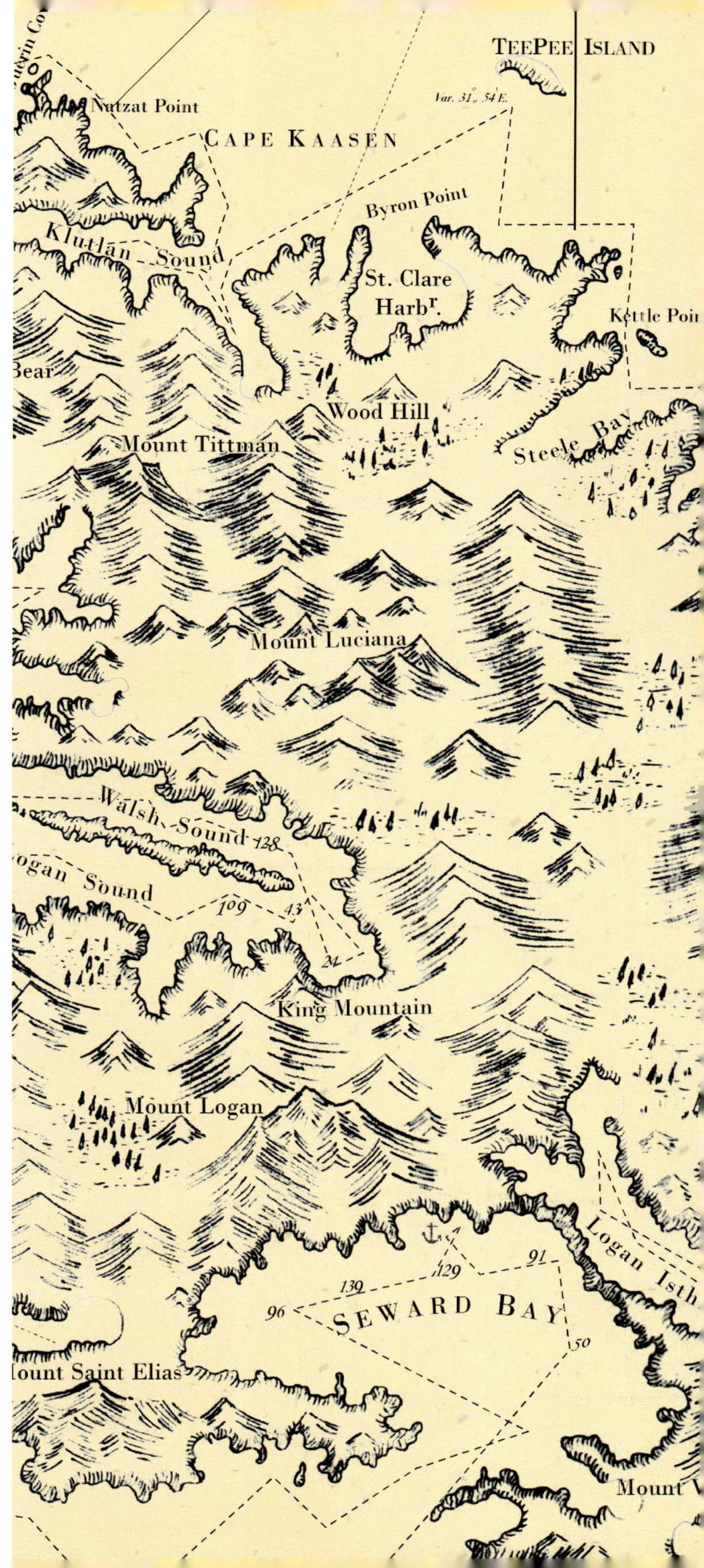

> *Intra-watershed relationships are central to China's struggle to urbanize amidst increasingly dire water shortages.*

Yellow River Basin »

Andreas Viglakis

This map of the Yellow River (*Huang He*) Basin appears in an atlas of China's major watersheds that offers an alternative reading of Chinese geography. By swapping out the country's political boundaries with a drainage basin, the map rearranges the traditional cartographic order to highlight hydrological connections within and across China's national borders.

A standard political map relates areas according to administrative boundaries, but this project reveals how areas shown to be unrelated on such maps are in fact closely bound together through shared natural systems. I employed two simple techniques at the outset: 1) masking areas outside each basin to focus the viewer's attention and 2) the 90-degree rotation of each map, which aligns the page to the west-east flow of China's major rivers. The rotation emphasizes each basin's dominant river network. I displayed settlements of all sizes at higher densities and sometimes at bolder weights than normal to help the viewer make connections between the places now linked by their relationship to the Yellow River and its watershed. Finally, I used special care in the background elevation layers, so that the contrast and color palette required to dramatically render the dialogue between river and land did not prevent the map from communicating the region's fundamental aridity.

Intra-watershed relationships are central to China's struggle to urbanize amid increasingly dire water shortages. These maps enable the viewer to see past the man-made divisions that commonly define national atlases and focus on a natural system vital to addressing the country's—and the region's—looming environmental problems.

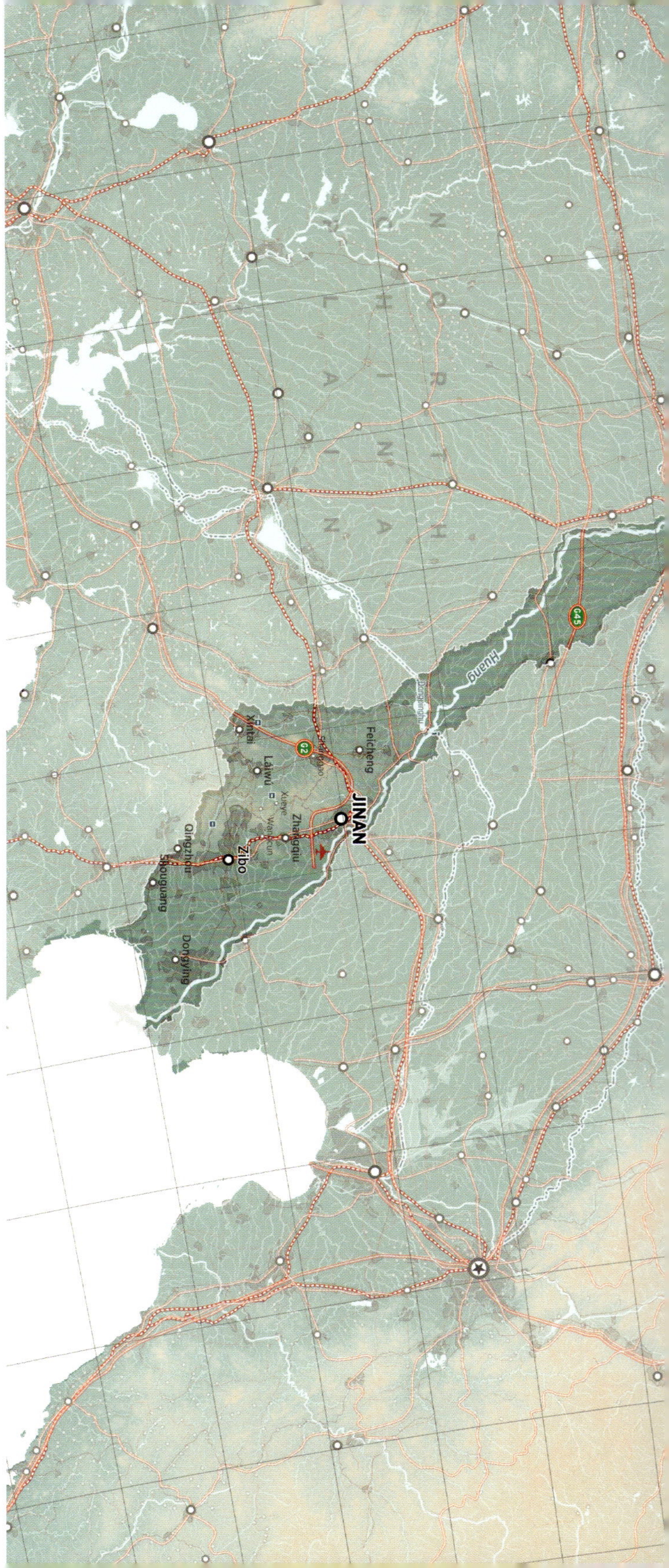

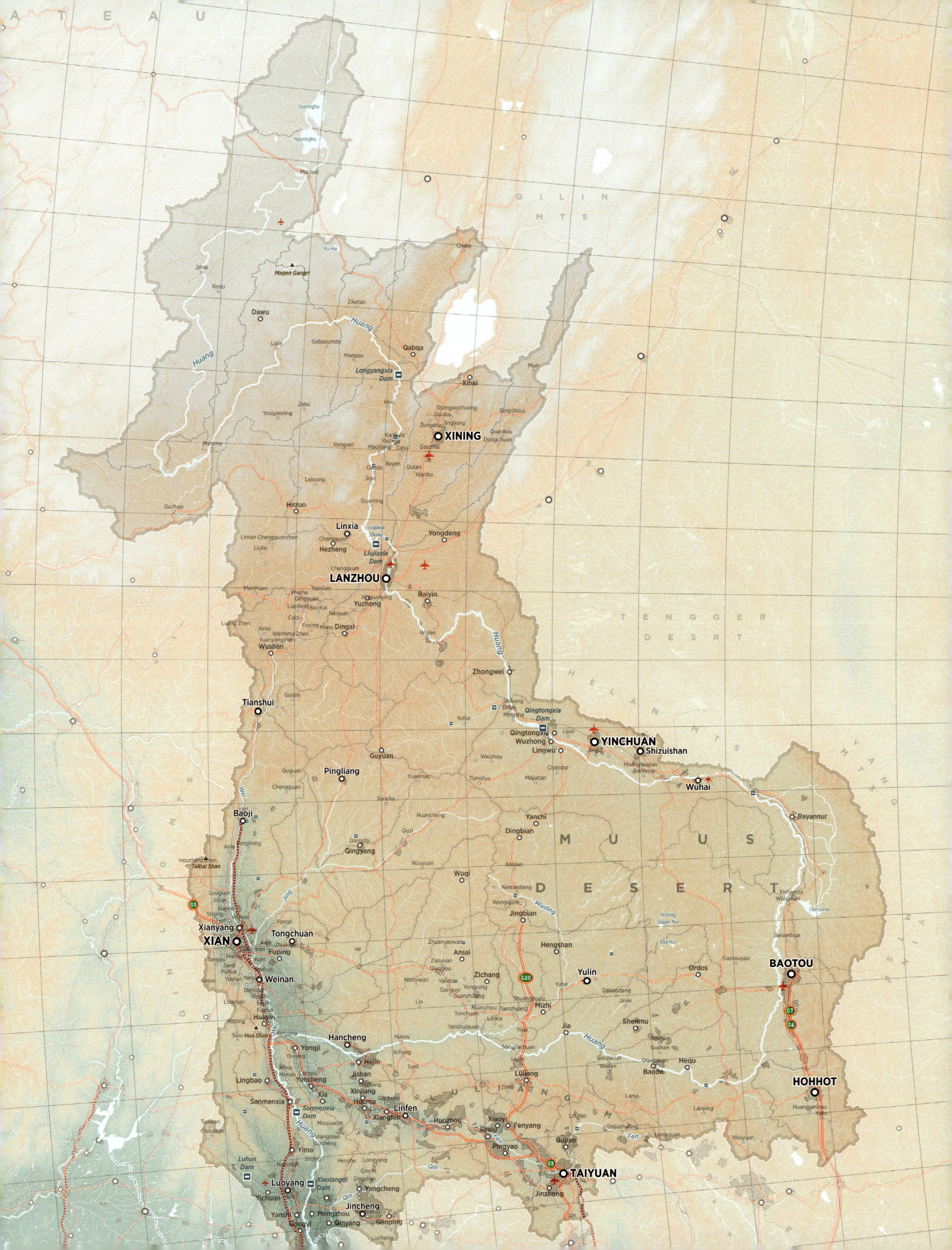
XINING
LANZHOU
YINCHUAN
Shizuishan
Wuhai
BAOTOU
HOHHOT
TAIYUAN
XIAN
Xianyang
Tongchuan
Weinan
Baoji
Tianshui
Pingliang
Guyuan
Qingyang
Zhongwei
Baiyin
Linxia
Hezhou
Hezuo
Dawu
Qabqa
Xinai
Yongdeng
Dingxi
Yuzhong
Wushan
Wuqi
Dingbian
Yanchi
Jingbian
Hengshan
Yulin
Ordos
Shenmu
Mizhi
Zichang
Ansai
Hancheng
Linfen
Houma
Xinjiang
Jishan
Hejin
Yongji
Yuncheng
Lingbao
Sanmenxia
Luoyang
Jincheng
Yangcheng
Huozhou
Fenyang
Jiexiu
Pingyao
Jinzhong
Lüliang
Baode
Hequ
Bayannur
Lingwu
Wuzhong
Qingtongxia
Qingtongxia Dam
Liujiaxia Dam
Longyangxia Dam
Sanmenxia Dam
Xiaolangdi Dam
Huang
Wei
Jing
Fen
Qin
Wuding
MU US
DESERT
TENGGER
DESRT
QILIN
MTS
HELAN MTS
ATEAU

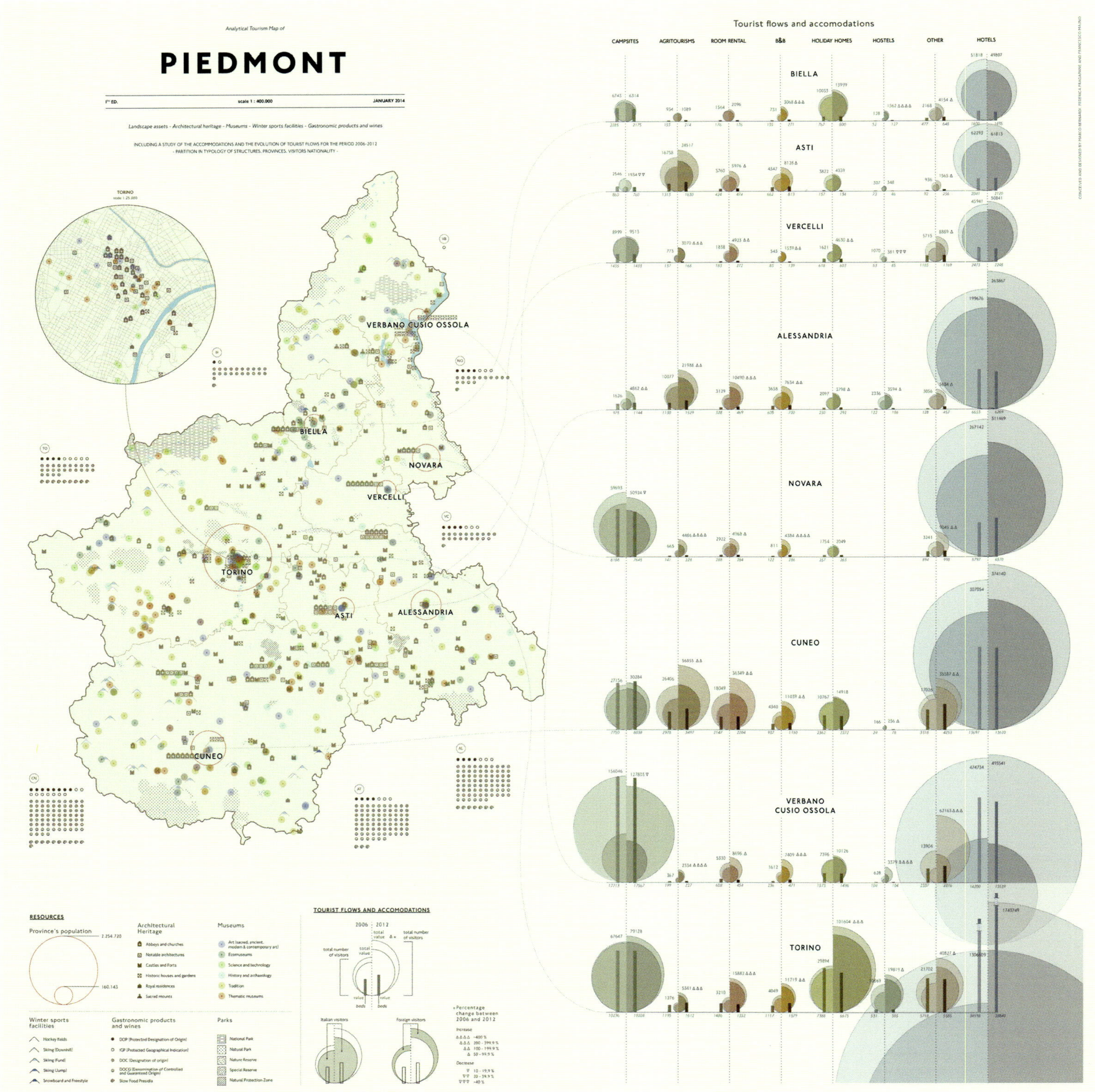
Analytical Tourism Map of
PIEDMONT
I^st ED.
scale 1 : 400.000
JANUARY 2014
Landscape assets - Architectural heritage - Museums - Winter sports facilities - Gastronomic products and wines
INCLUDING A STUDY OF THE ACCOMMODATIONS AND THE EVOLUTION OF TOURIST FLOWS FOR THE PERIOD 2006-2012
- PARTITION IN TYPOLOGY OF STRUCTURES, PROVINCES, VISITORS NATIONALITY -
TORINO
scale 1:25.000
VERBANO CUSIO OSSOLA
BIELLA
NOVARA
VERCELLI
TORINO
ASTI
ALESSANDRIA
CUNEO
Tourist flows and accomodations
CAMPSITES
AGRITOURISMS
ROOM RENTAL
B&B
HOLIDAY HOMES
HOSTELS
OTHER
HOTELS
BIELLA
ASTI
VERCELLI
ALESSANDRIA
NOVARA
CUNEO
VERBANO CUSIO OSSOLA
TORINO
RESOURCES
Province's population
2.254.720
160.143
Architectural Heritage
Abbeys and churches
Notable architectures
Castles and Forts
Historic houses and gardens
Royal residences
Sacred mounts
Museums
Art (sacred, ancient, modern & contemporary art)
Ecomuseums
Science and technology
History and archaeology
Tradition
Thematic museums
Winter sports facilities
Hockey fields
Skiing (Downhill)
Skiing (Fund)
Skiing (Jump)
Snowboard and Freestyle
Gastronomic products and wines
DOP (Protected Designation of Origin)
IGP (Protected Geographical Indication)
DOC (Designation of origin)
DOCG (Denomination of Controlled and Guaranteed Origin)
Slow Food Presidia
Parks
National Park
Natural Park
Nature Reserve
Special Reserve
Natural Protection Zone
TOURIST FLOWS AND ACCOMODATIONS
2006 2012
total number of visitors
total value
total value
total number of visitors
value
beds
value
beds
Italian visitors
Foreign visitors
*Percentage change between 2006 and 2012
Increase
ΔΔΔΔ >400 %
ΔΔΔ 200 - 399.9 %
ΔΔ 100 - 199.9 %
Δ 50 - 99.9 %
Decrease
∇ 10 - 19.9 %
∇∇ 20 - 39.9 %
∇∇∇ >40 %

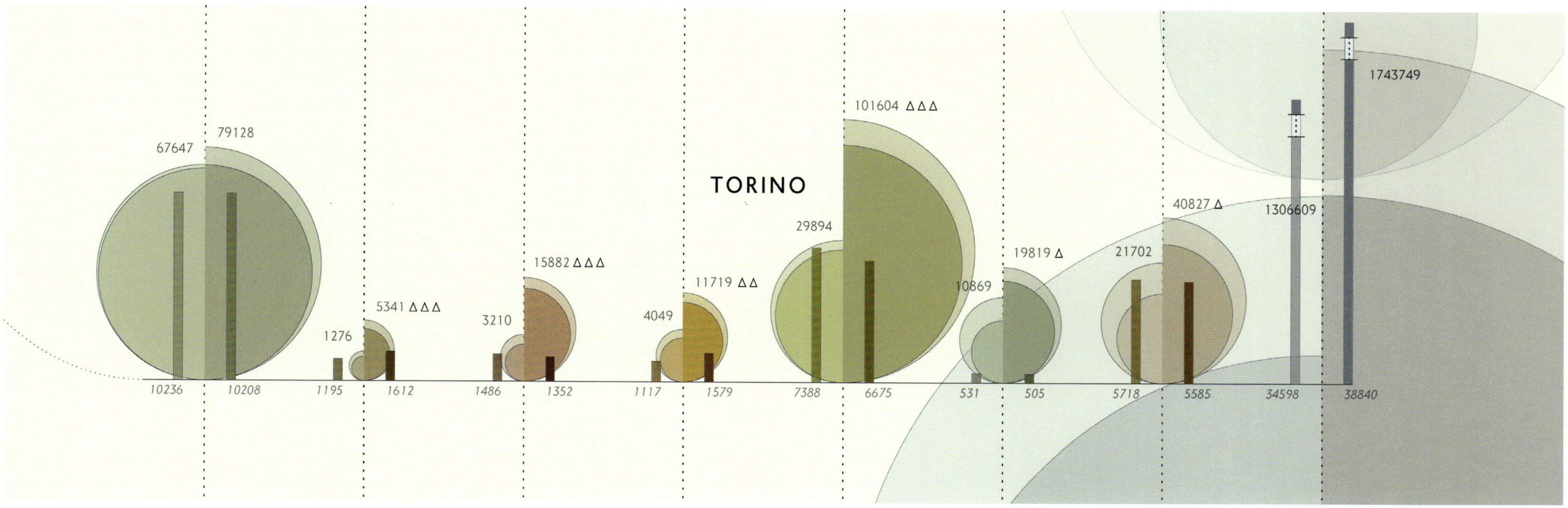

« The Analytical Tourism Map of Piedmont

Marco Bernardi, Federica Fragapane, Francesco Majno

This map has been realized by analyzing the Open Data made available by the Piedmont region in Italy. The work investigates tourism in the area in a complete and detailed way, considering the attraction factors, the evolution and dimension of tourist flows, and tourist accommodations, including hotels, bed & breakfasts, campsites, rented rooms and homes, and hostels.

In designing the map, we had to focus on multiple levels of information because we didn't want to eliminate anything. Our challenge was to push the boundaries of how different languages and elements could coexist in one single, extremely rich artwork without compromising legibility. To achieve this goal, the background had to be light and simple, showing just province boundaries, lakes, and major waterways. We drew nature parks with patterns, while winter sports facilities are represented by tiny mountain icons, symbolizing altitude. We designed small icons for most important architectural heritage buildings, including abbeys, castles, and historic houses. Finally, a layer of circles plotted on the map shows important museums, colored according to theme.

Geolocation of everything on the map would have made it too crowded, so we started to surround the map with additional data, designing custom abstract icons for food products and wines. We also designed a detailed visualization of accommodation facilities and tourist flows, clearly linked to the map with curvy lines.

The final result is a well-balanced mix of visual languages and unique elements that readers can explore over and over, always finding something new and getting lost in the details.

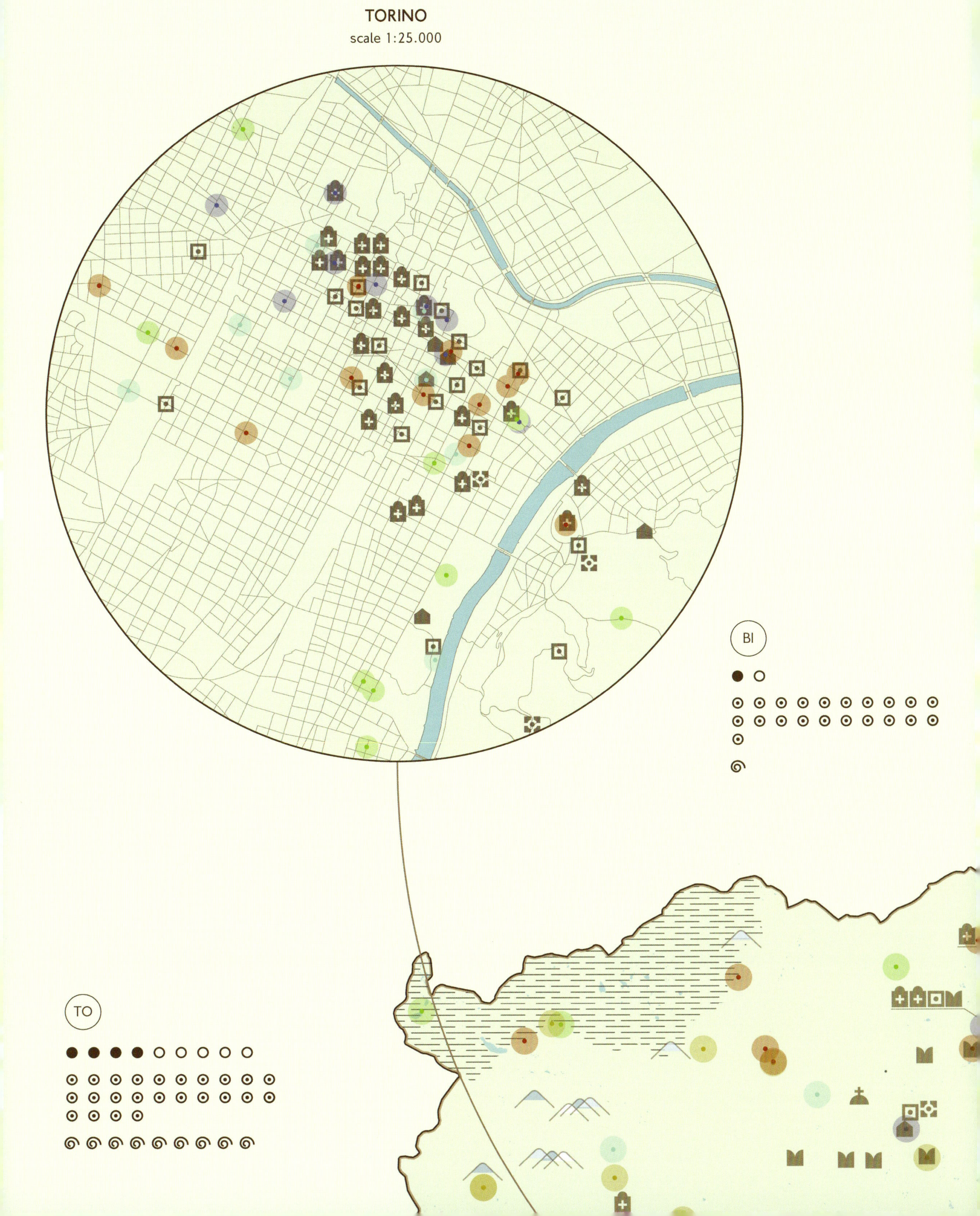

TORINO
scale 1:25.000
BI
TO

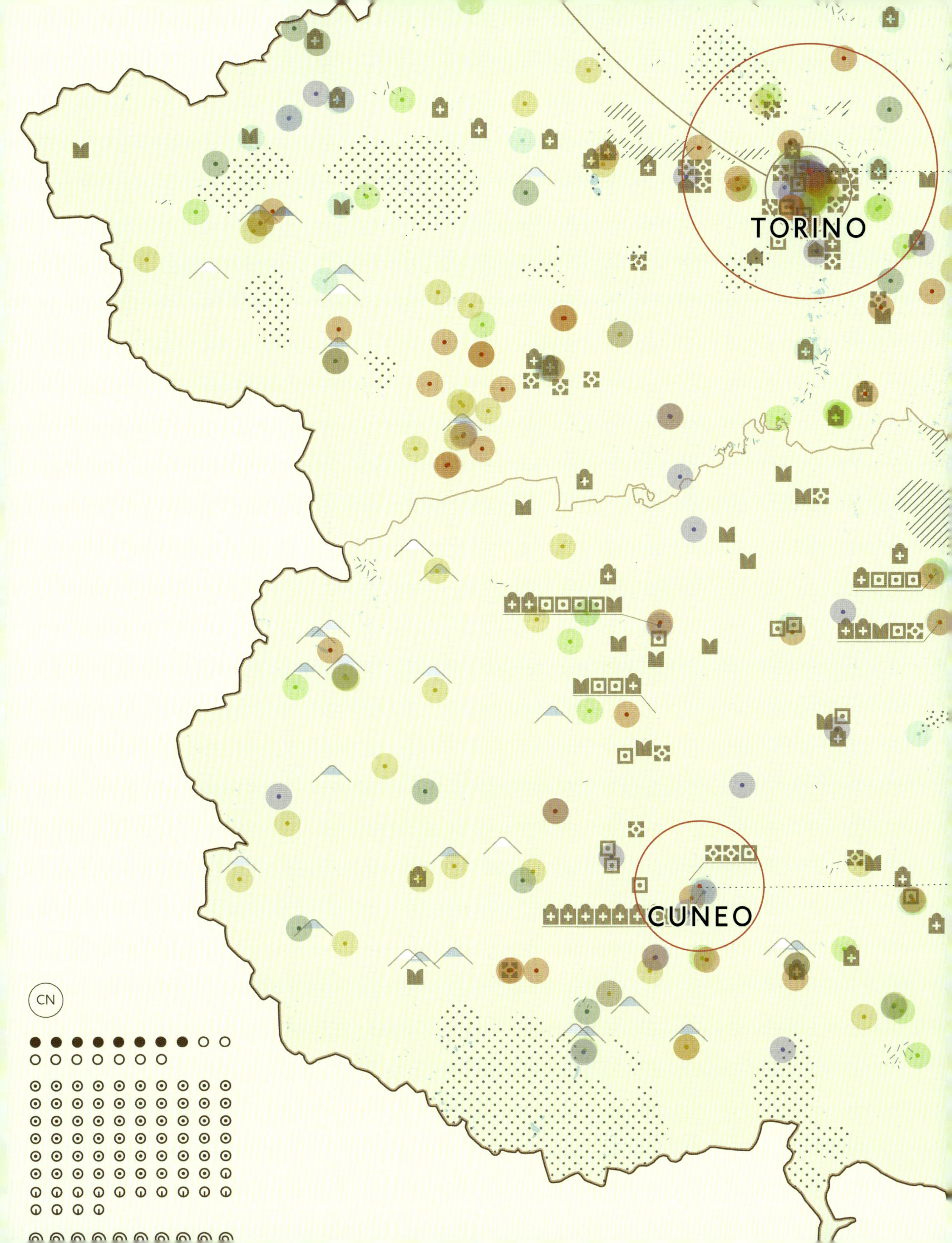
TORINO
CUNEO
CN

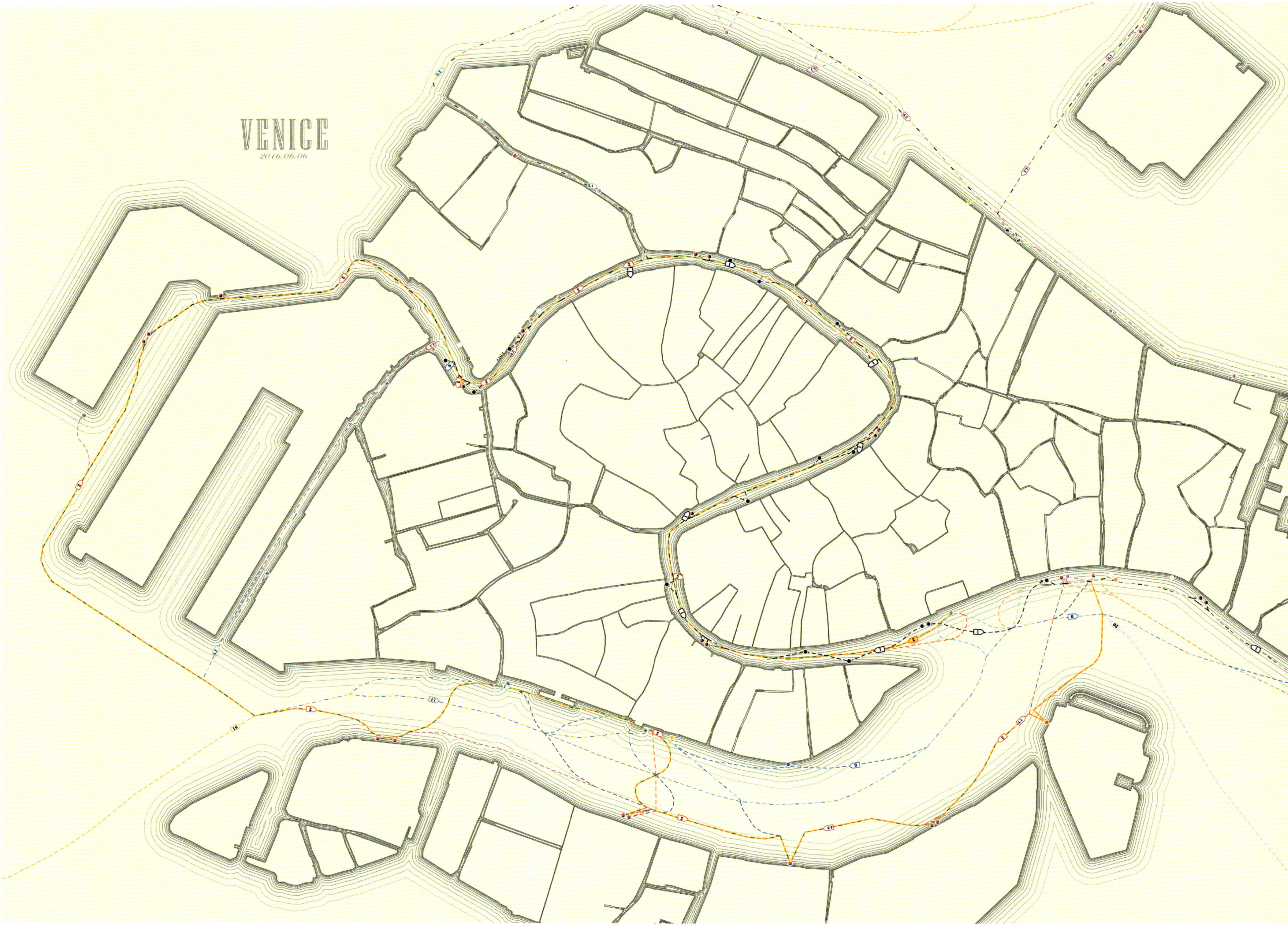
VENICE
2016.06.06

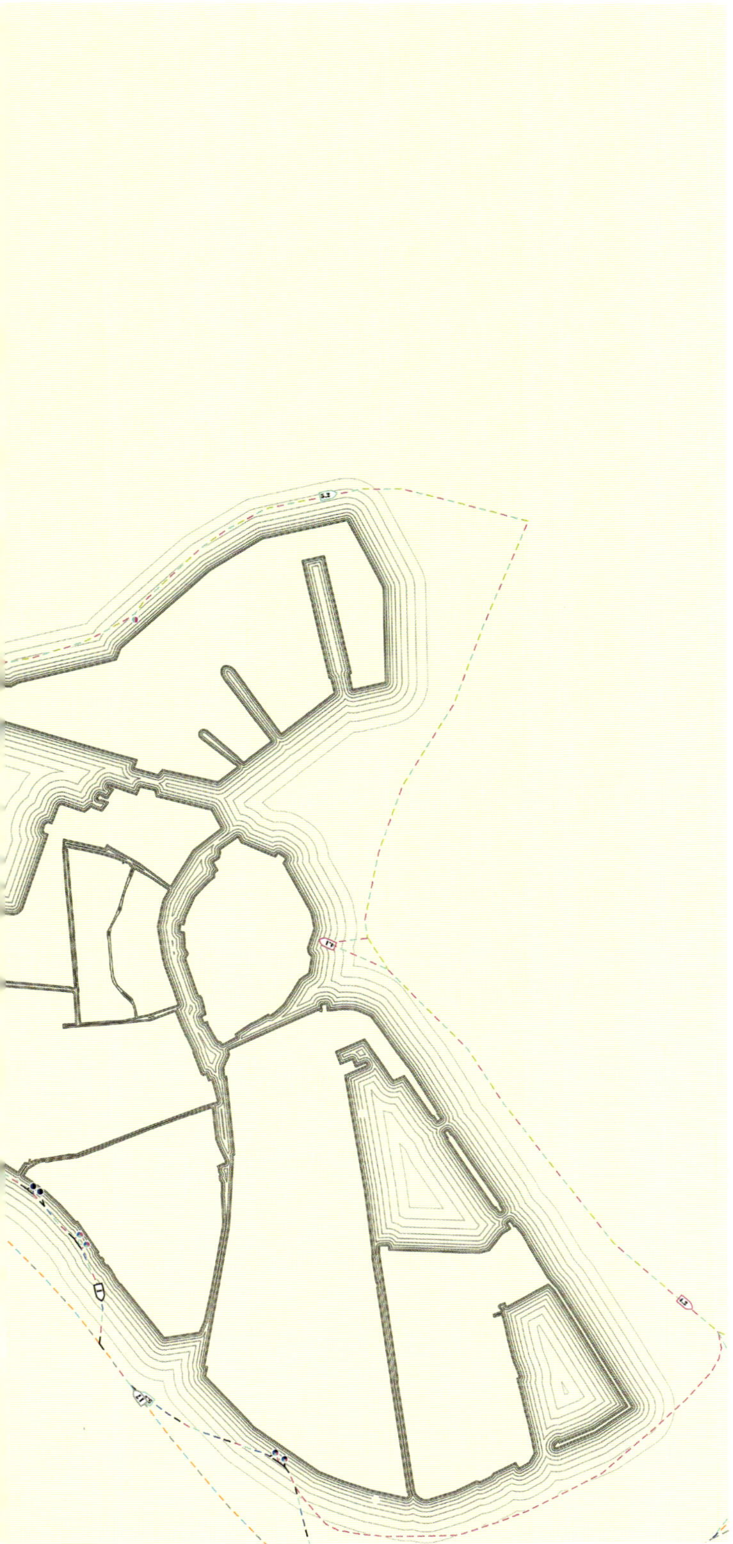

« Venice Vaporetto Map

Edvardas Ščerbavičius

A *vaporetto* is a public water bus that transports passengers through the canals of Venice, Italy. This map depicts the vaporetto network and its real-time boat traffic. I've created a modern web map that lets viewers zoom in and pan around, select and highlight individual stops and boats, and view route timetables.

Because Venice is such an iconic, historic, island city, I wanted to give the map an antique look and feel. After studying many old maps, I was drawn to the aesthetic of seventeenth-nineteenth-century Western cartography. In that era, mapmakers used the waterlining technique—lines drawn parallel to the edge of a water feature, decreasing in proximity and strength farther from the edge. I used the waterlining technique for my base map, which helped to make a clear distinction between sea and land without introducing additional patterns or colors.

To create a visual hierarchy I used monochromatic texture and shades of grey for waterlines in the background.

The vaporetto network itself is represented by dashed lines. To show places where many different routes follow the same path I used a pattern of multi-colored dashes, with each color representing an individual route. A similar design approach is used for the stops. Circles depict places the water bus stops and are divided into equal slices, each colored according to the vaporetto routes that visit the stop.

Finally, to help visitors to Venice who are using the map to navigate the city, colors of the waterbus routes and boats match the actual colors used on vaporetto station signs.

The Historical Ecology of Upper San Francisquito Creek »

Erik Steiner

I am particularly fond of the idea of *landscape*. In a basic sense, it simply describes one's field of view, what you can see from a certain vantage point. In a deeper sense, landscape embodies a scale of appreciation, a method of interpretation, and a philosophy of understanding.

Painters and photographers know landscape well, as their medium all but requires a singular vantage point. But cartographers have a beef with perspective. Afterall, Ptolemy, Mercator, and others invented "the view from nowhere" in order to depict Earth on a flat page. Pictorial maps and bird's-eye views notwithstanding, the top-down view from the sky has clearly dominated since the Enlightenment. And now, current mapmaking methods make any other vantage point too particular, not scalable, or easily automated. Furthermore, critical cartographers observe the inherent subjectivity of an author's vantage point, and this cuts against our fundamental ambition for accuracy in mapping.

Landscape cartography shares more with landscape photography than modern mapmaking. It operates at the human scale, with a focus on the visible, but with a holistic appreciation for how a place functions. Geographer Carl Sauer describes landscape as embracing "the whole dynamic relation of life and land."

My goal with this map was to use cartography to understand a landscape's past and help to imagine its future. Since the arrival of European settlers about 250 years ago, the San Francisco Peninsula been shaped by dramatic change. The area depicted here—the headwaters of the San Francisquito Creek above Silicon Valley—is no exception, despite the fact that it has been part of an exclusive biological preserve since 1971.

This map is a combination of two views separated by two centuries, their juxtaposition an inquiry into the complex and often subtle changes affected by grazing, logging, urbanization, fire, climate change – and most visibly by the 1892 Searsville Dam and its defunct, sediment-filled reservoir. These views are both simulations, but their careful rendering honors the complexity of their ecosystems and hopes to inspire the viewer to read the land more than read the map.

What will become of this dam? And what will happen to the landscape around it? Will it eventually return to its "original" state before the arrival of European settlers?

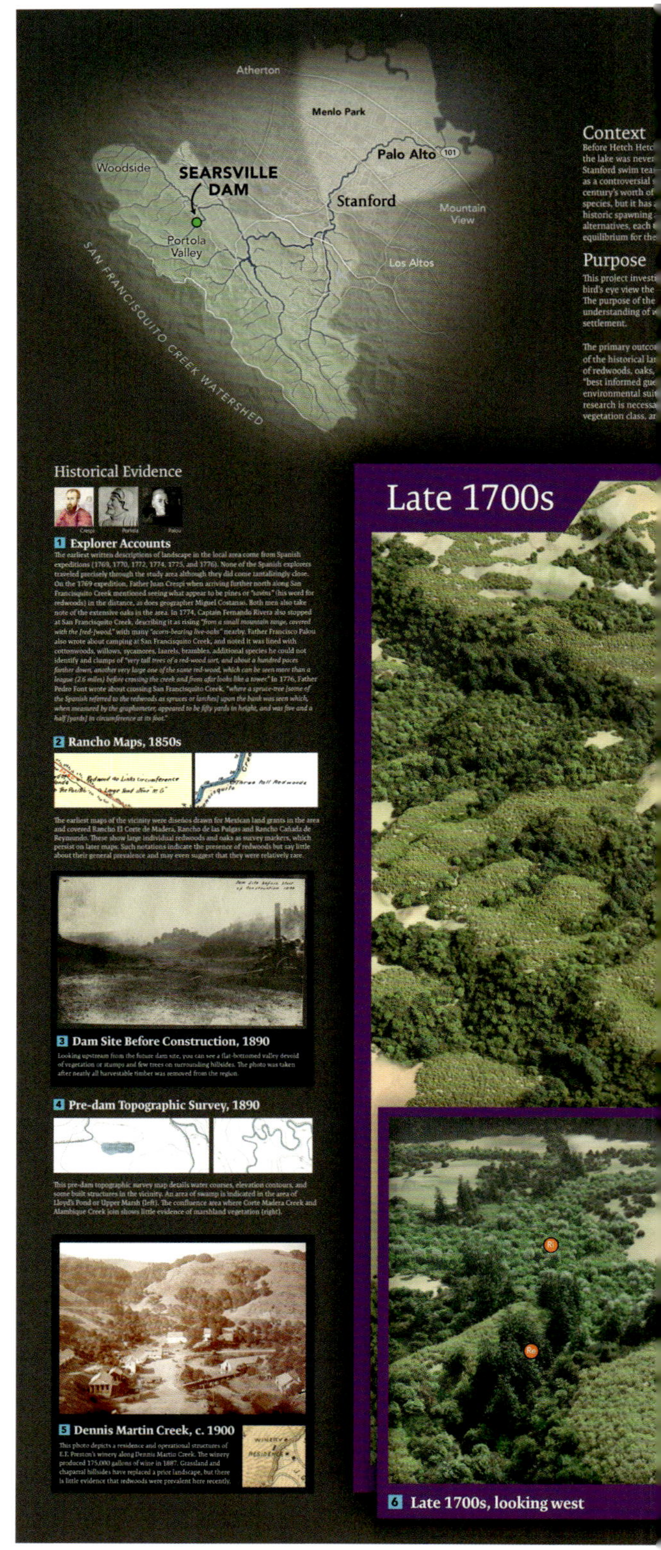

…Dam. Built in 1892 to supply the burgeoning city of San Francisco, …r supply. Instead, it became a recreation destination used by the …t of the Jasper Ridge Biological Preserve. The aging dam stands today …ked to the ecosystems in and around the San Francisquito Creek. A …s helped create a rich upstream wetland habitat for bat and bird …ervoir and the dam blocks steelhead trout from returning to their …ity is currently undergoing a study to evaluate Searsville Dam …ant environmental restoration to help establish a sustainable

…**logy** of the San Francisquito Creek watershed and reconstructs a …e looked before the area was settled by Europeans (circa late 1700s). …rmation for future restoration discussions, to improve our …e, and how the ecosystem functioned prior to major human

…dscape reconstruction - a **visual synthesis** of our best understanding …torical evidence, expert advice, and GIS analysis. The precise extents …asses are exceedingly difficult to determine. The maps here show a …nt, size, and density of these vegetation classes based on …ned by historical assumptions (described on the far right). Further …l assumptions, make informed adjustments to the extents of each …ction of historical conditions in specific areas of interest.

The Historical Ecology of Upper San Francisquito Creek

A landscape reconstruction that shows what the area around Searsville Reservoir may have looked like in the late 1700s.

Legend and Historical Assumptions

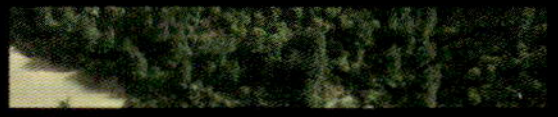

Re Redwood Forest

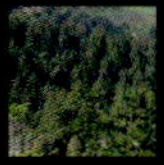

Explorer descriptions of the region 1 frequently noted the size, quality, and abundance of redwoods in the Santa Cruz Mountains. Their journals also indicate there were several isolated stands, including below the site of the present-day dam along San Francisquito Creek extending down to the future site of Palo Alto. Later Rancho maps 2 used redwoods as survey markers, indicating their presence but not their prevalence.

The transition from redwood-dominant forest to oak-chaparral-grassland vegetation types happened within the study area, although it is unclear where. There is some evidence that historic climate conditions were slightly cooler and wetter than today, with increased fogginess supportive of redwood growth.

Logging in the area began as early as 1777 with the establishment of Santa Clara Mission near San Jose and continued through the Mexican Period. Eventually, with the growth of San Francisco during the period of 1840s-1870s, logging all but eliminated redwoods from the study area and in much of the Santa Cruz Mountains 3.

Redwoods and other conifers present in the area today are second-growth stands less than 150 years old found near streams, on private property, and in the upper stretches of Dennis Martin Creek. A few isolated redwood trees are present below the dam.

A photograph 5 looking up Dennis Martin Creek around 1900 shows no apparent evidence of redwood stumps, and mostly barren steep hillsides that have since been replaced by oak woodland.

Species Composition
Coast Redwood
Douglas Fir
Tan Oak
Madrone

Conditions
Cool and stable temperature regimes
Moist, sheltered areas require coastal fog
North-facing slopes

O Oak Woodland and Savanna

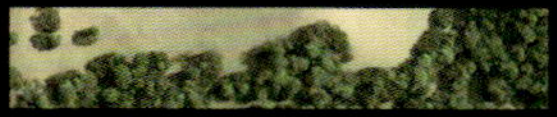

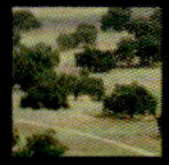

Due to their high nutritional value and abundance, acorns of the tanbark, blue, live and valley oak were a significant source of food for Native inhabitants of this area. Native grasses and other seed bearing plants were also important. Their land management practices would have been oriented toward supporting nut and seed production. Most significantly, these included burning woodland undergrowth to thin the woodland and understory shrubs, improve visibility, enable access, and to promote seedling growth.

With the arrival of the missions, oaks were a significant source of timber and were heavily logged throughout the 1800s, although less so than redwoods. The introduction of substantial grazing would have suppressed oak regeneration in the grassland communities and adjacent areas, sustaining the open oak savanna that had been previously maintained by fire.

In the last century with the absence of fire, logging, and (more recently) grazing, oaks have regrown to cover historical extents in the Jasper Ridge Biological Preserve, residential areas, and in areas where there were previously redwood stands. Although it is difficult to determine the exact extent of oaks historically, it is assumed that they would have required similar site conditions as today's forest.

However, oak woodlands today are likely more homogenous with denser undergrowth than their historical structure. The historic woodlands and savanna likely had a more open canopy and varied tree ages, notably with older, larger trees. In open savanna areas, there were likely fewer trees overall as secondary shrub and seedling recruitment have filled out the prairie in the last century.

Species Composition
Coast Live Oak
Valley Oak
Blue Oak
Leather Oak
California Bay
California Buckeye
Black Oak
Madrone
Toyon
Prunus (plums and cherries)

Conditions
Dry, well drained areas
Mild winters and hot summers

Ri Riparian Forest

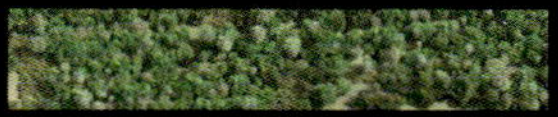

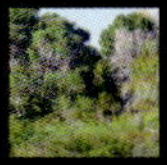

A detailed topographic survey of the study area performed before dam construction began in 1891 4 reveals some clues that the historical streams were mostly free-flowing with moderate gradient profiles 8.

Sediment deposition over centuries led to the establishment of a flat-bottomed canyon under the area of the current lake extending up Corte Madera Creek. These meandering channels would have likely supported a wide riparian zone along the current and historical stream tributaries.

Further upstream, Sausal Creek likely supported a broad riparian forest in flatter areas along the San Andreas Fault above where it joins the steeper Dennis Martin channel. Dennis Martin Creek, with its steep profile, would have had a narrow riparian belt but this would have spread out as the streams joined Alambique Creek.

Substantial erosion as a result of deforestation upstream and subsequent sedimentation since the dam construction has lowered all of the stream water profiles and promoted riparian and marshland vegetation, particularly along Alambique Creek and its confluence with Corte Madera.

Historically, the riparian forest structure would have been more varied with a more open canopy due to beaver activity and fire. The areas adjacent to streams were almost certainly logged over and used for irrigation and grazing 3 in the 1800s. The recovered riparian forest is younger and dominated by moisture-tolerant species than it was historically.

Species Composition
Willow
Box Elder
California Bay
Buckeye
Alder
Elderberry
Big leaf Maple
Cottonwood

Conditions
Gently sloped valleys adjacent to streams
Moist conditions and buffered temperature

Gr Grassland

Historically, native grasslands were abundant in the study area, maintained through frequent burning from lightning- and human-caused fires and grazing by natural herbivores.

Native Americans burned grasslands and chaparral to enable mobility, encourage grass and seed production, drive game, and promote visibility. Frequent burning could have limited the extent of sagebrush and chaparral.

Significant grazing was introduced with ranching during the Mission and Mexican periods and later. Grazing suppressed chaparral and oak regrowth and favored the non-native grasses introduced by the Spanish. The area was also cultivated as grain fields, vineyards 3, and orchards, which were developed on existing grassland or cleared chaparral. In combination with the significant logging during the 1800s, by the time of the dam construction the area was mostly devoid of trees 3.

Livestock were active in Jasper Ridge until 1960 and in surrounding areas for longer. In the last half-century, especially with fire-suppression and the elimination of logging and grazing, oak woodlands, savanna, and chaparral have reclaimed significant parts of the landscape, reducing the extent of grasslands.

Areas adjacent to Searsville Lake and the tributaries of San Francisquito Creek have substantially recovered as dense wooded areas. Although non-native European grasses have replaced most of the native grasses, some areas still persist, principally on the high serpentine ridge to the east of the study area.

Species Composition
Purple needle grass
Slender hair grass
Little California Melica
(now replaced with non-natives)

Conditions
Wet and cool winters, dry and hot summers
Well-drained flat plains to gently rolling foothills

Ch Chaparral

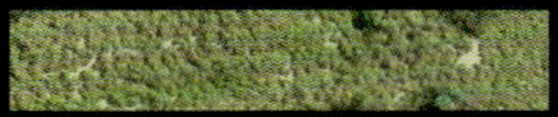

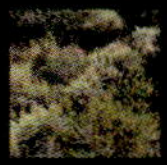

In dense, woody stands, chaparral was generally undesirable for Native inhabitants who burned mature stands regularly to encourage herbs and forbs, improve mobility and support hunting. As chaparral is extremely flammable, these fires may have been difficult to control and could have charred broad areas before recovering from deep roots within 5-10 years. Areas that were burned repeatedly in short cycles could have been converted to grasslands.

Livestock grazing introduced in the 1800s would have contributed to keeping chaparral at bay on the edges of pastureland. Subsequently, chaparral would have been painstakingly removed to allow for cultivation of grain, vineyards and orchards. Chaparral and its roots could be sold for wood and charcoal, especially during the late 19th century boom.

In the last century with fire suppression and the reduction of grazing, chaparral has recovered in areas that were grazed or logged. Some areas may have been chaparral continuously since the 1700s as large uninhabitable areas are still noted on Stanford property in the late 1800s.

With such a varied history, it is difficult to assess the exact extent of historical chaparral at any given time compared to today, we can only presume similar site conditions and distribution.

Species Composition
Chamise
Manzanita
Buck Bush
California Sage
Holly leafed Cherry
Coyote Bush
Coffeeberry
Redberry
Toyon
Yerba Santa

Conditions
Dry, west- and south-facing slopes, steep slopes and ridges above drainages

M Marshland

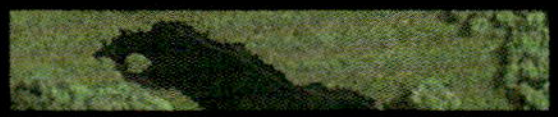

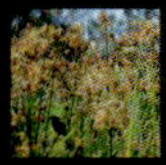

With respect to ecological changes in the study area over the last two centuries, the growth of ponds and marshland is perhaps the most apparent (aside from the Reservoir itself). Substantial sedimentation from Sausal, Dennis Martin, and Alambique Creeks above Searsville Lake has significantly modified the topography, transforming their stream profiles from moderate to very low 8. Their confluence now forms a wide, flat, marshy area surrounded by riparian forest.

The current wetland vegetation supports several aquatic and semi-aquatic species in a broad zone extending up to and beyond Portola Road to Upper Marsh. With strong seasonal stream flow patterns, these areas are typically dry during the summer and fall and moist during the winter and spring.

From historical maps we know of two ponds or marshes in the area, the largest being approximately the area west of Portola Road denoted as "Lloyd's Pond." 4 The marsh was drained by Alambique Creek and was adjacent to the town of Searsville, established in the 1860s but ultimately abandoned when the reservoir was filled.

Few if no other areas would have likely supported marshland communities, even at the major confluence of Alambique and Corte Madera Creeks at the end of the current reservoir.

Species Composition
Tule
Cattail

Conditions
Wet, poorly drained areas fed by streams
All exposures and slopes if a depression is saturated or temporarily flooded

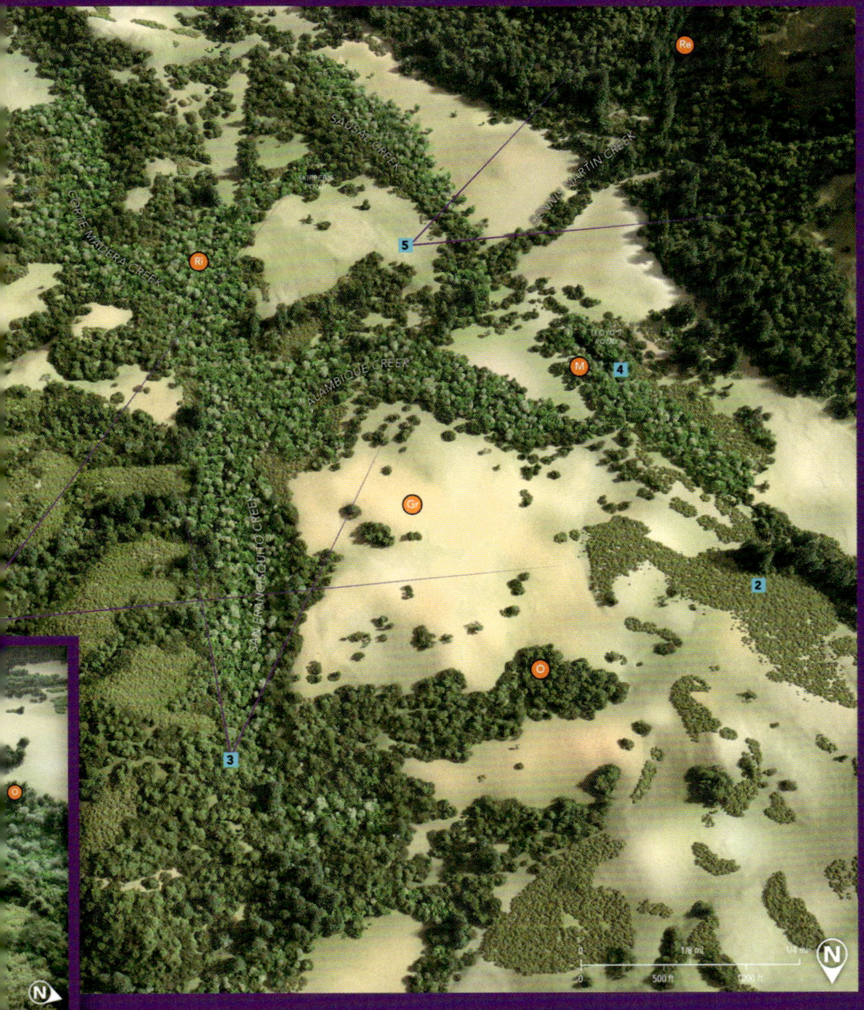

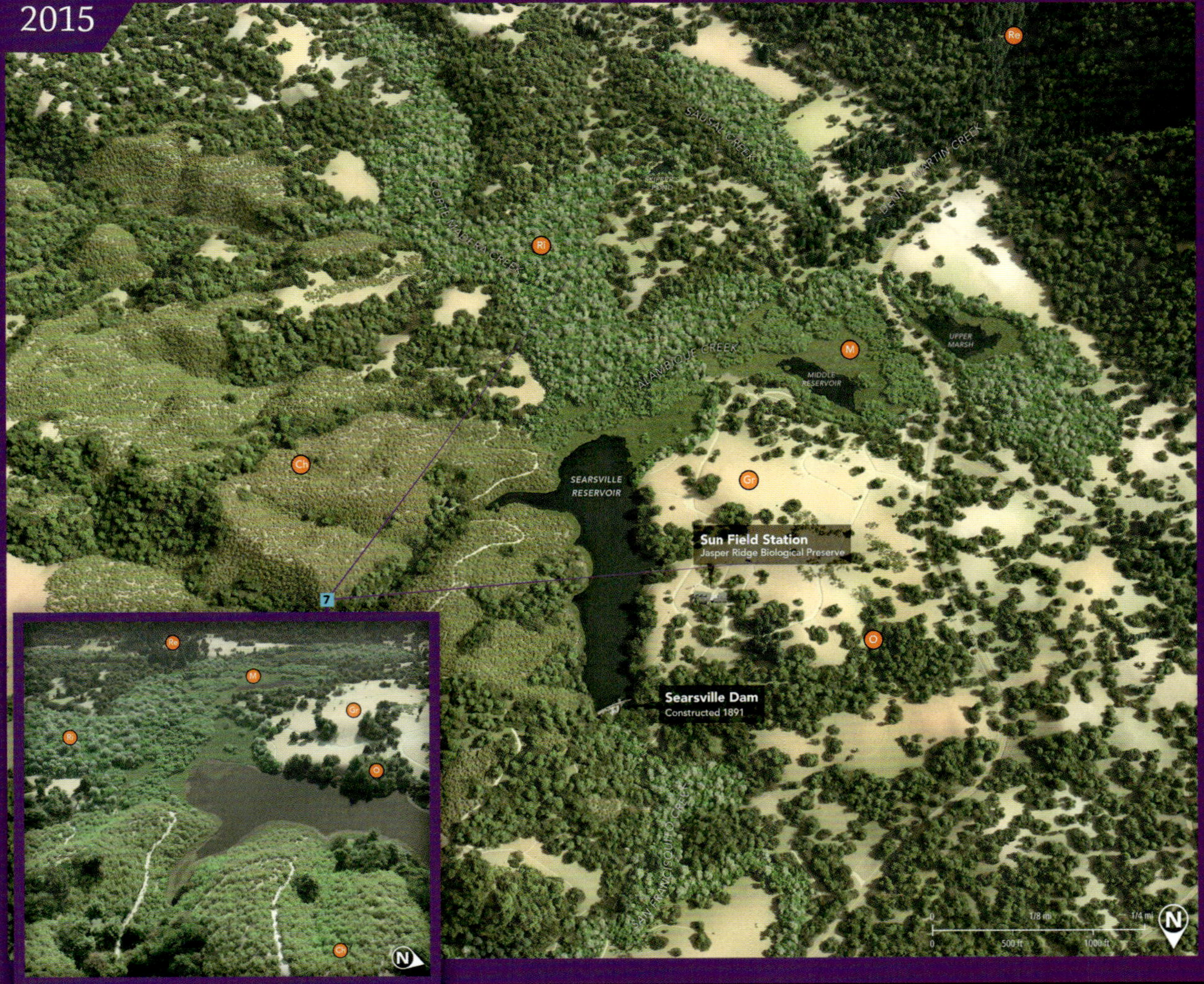

7 2015, looking west

Erik Steiner, Spatial History Project, Center for Spatial and Textual Analysis (CESTA), Stanford University
with contributions from Laura Jones, Matthew Walter, Julie Cain and Maria Joao Santos and many others

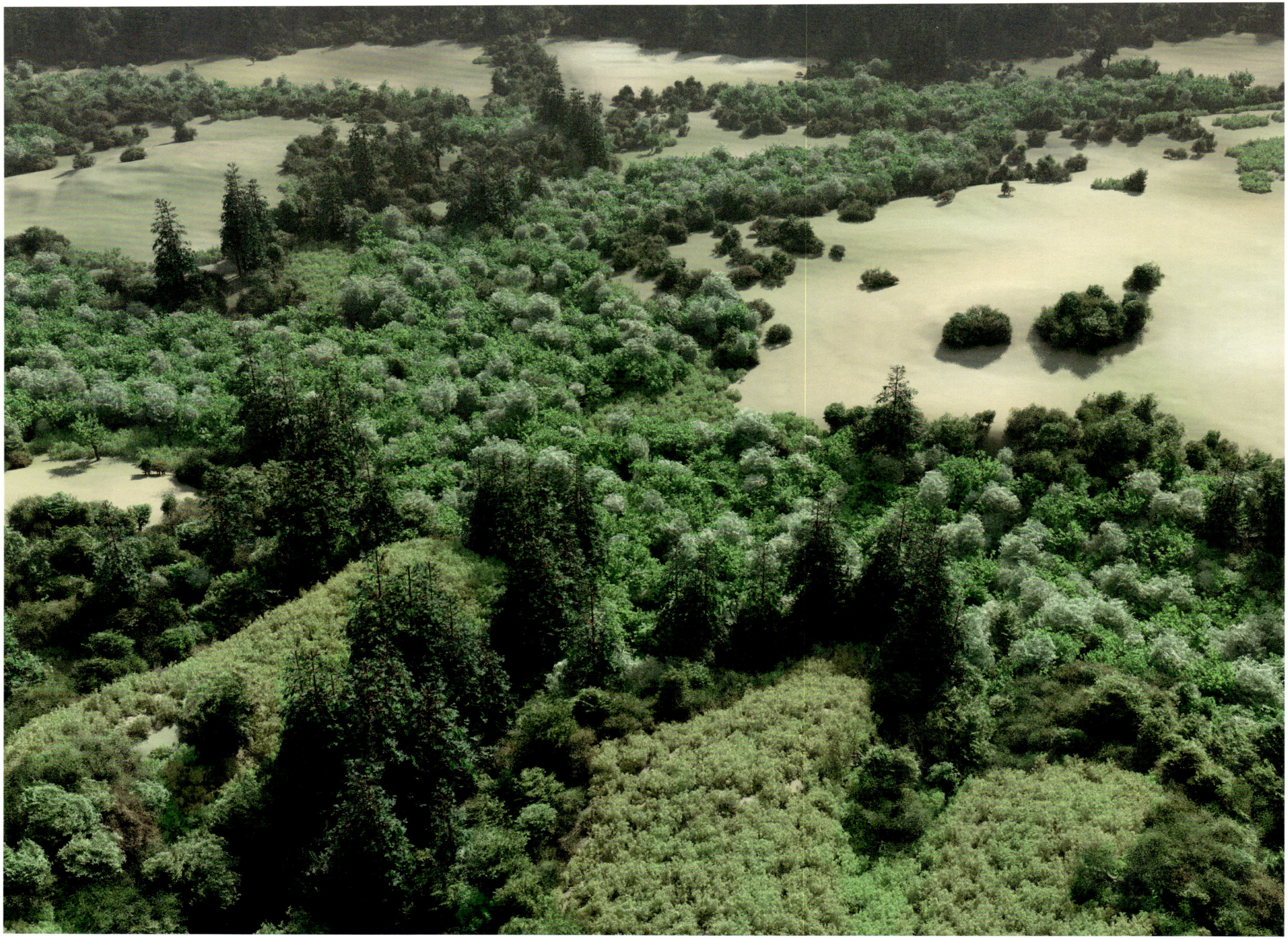

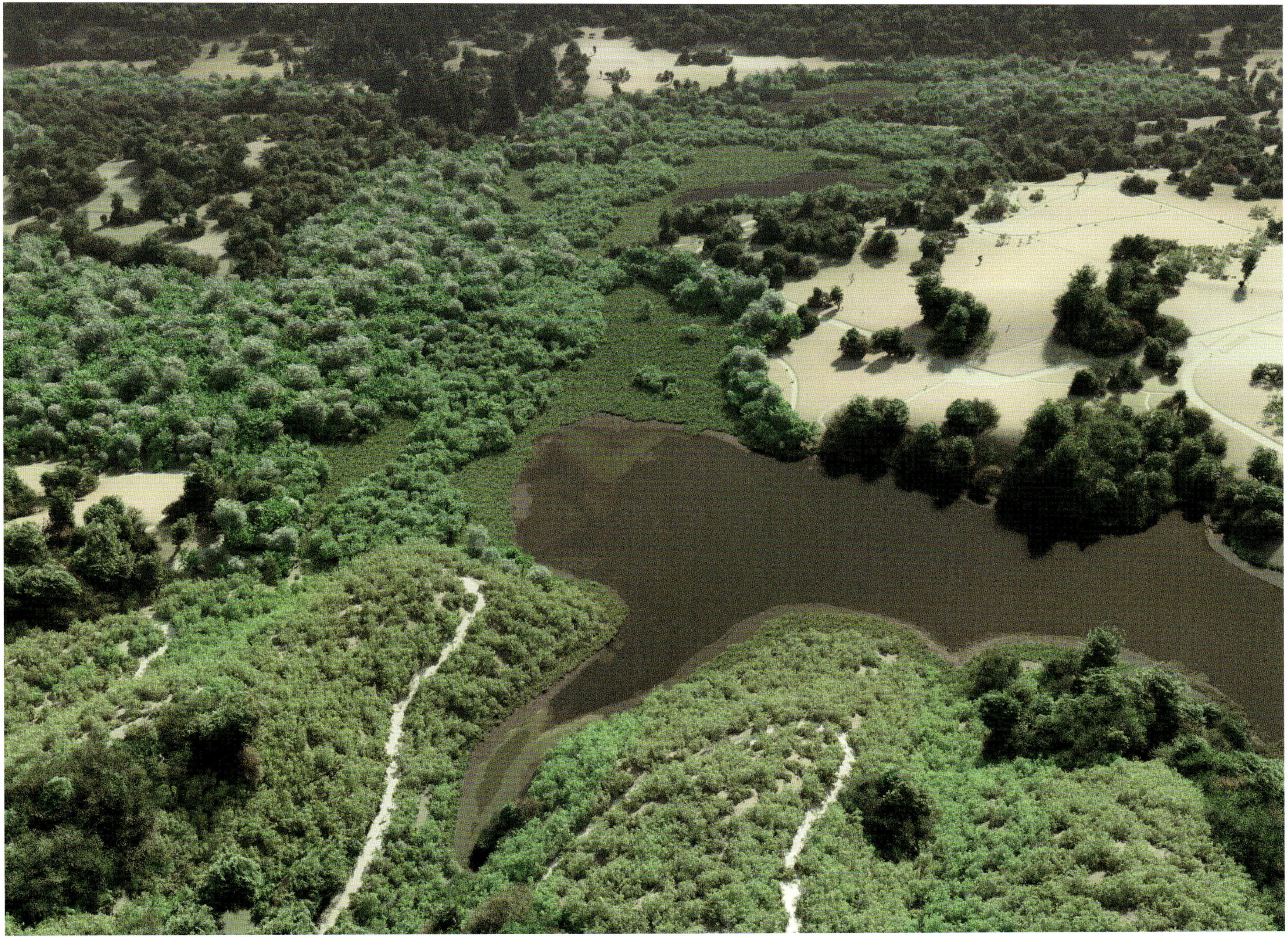

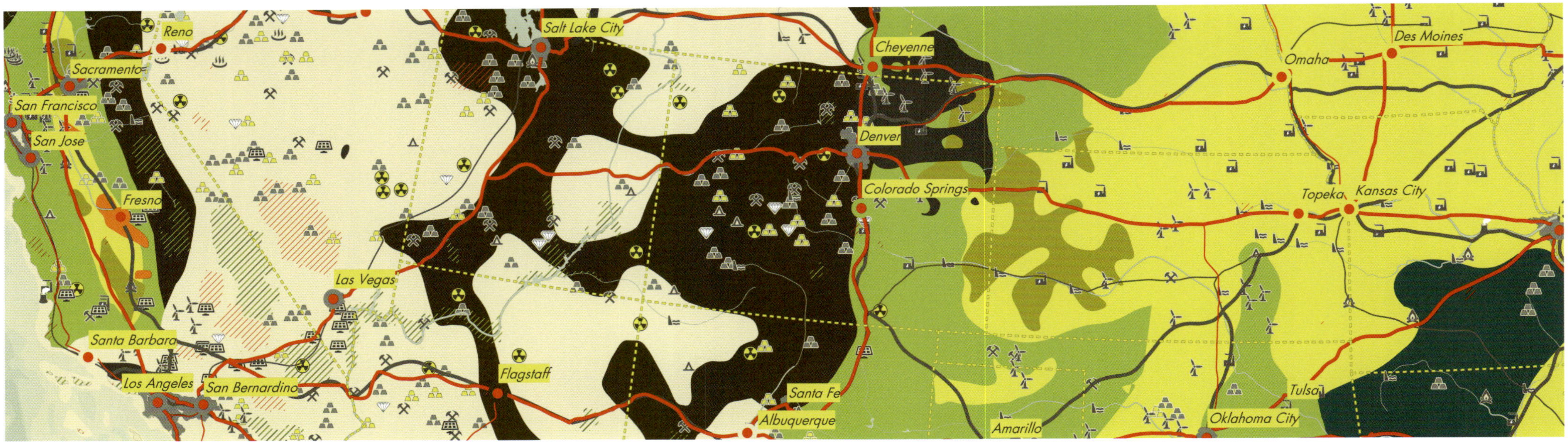

The United States: Her Natural and Industrial Resources »

Stephen Smith (TheMapSmith)

This map is a modern homage to a retro travel poster produced by the British Information Services in the early 1940s. The map depicts the natural and industrial resources of the contiguous United States overlaid on a simplified land use model. I pulled the color scheme of the map directly from the original, but implemented it differently due to differences in geography. My goal was to create something that would remind viewers of a grade school wall map and cause them to question whether they were looking at an old or new image.

Inspiration comes from many places, and I often turn to designs of yesteryear to guide my work. The creation of this map was an exercise in replicating the handmade appearance of the old map using modern data and tools. The environment in which you make your decisions greatly affects the outcome. For example, if you don't have the "Paint Bucket" tool of a graphics program, you might think differently about a design with filled shapes. If you don't have hundreds of typefaces to choose from, you might spend more time on the content of your design than the fonts. Constraints lead to creativity. By closely emulating historical designs you can replicate the constraints of the manual design era and often end up creating beautiful cartography.

The biggest challenge in creating this map was the smoothly generalized land use data. The United States Department of Agriculture's Cropscape Data Layer raster dataset was resampled to 15 kilometers using a majority rule, then converted to vector polygons and simplified. To limit the colors required in the final design, I merged some classes. Finally, I downloaded polygons for coal and oil plays from the Energy Information Agency. I then topologically aligned these datasets to the national boundaries.

The mineral resource points came from the USGS Mineral Resource Data System. I merged most of the natural resources into larger groups, but uranium was so abundant I left it in its own class. I grouped points so all locations within one kilometer are represented by a single point. Renewable energy source data came from the USGS and from OpenEI. Like the mineral resource data, points within one kilometer were grouped and represented by a single point. To represent hydropower, only dams near rivers and lakes already shown on the map were included.

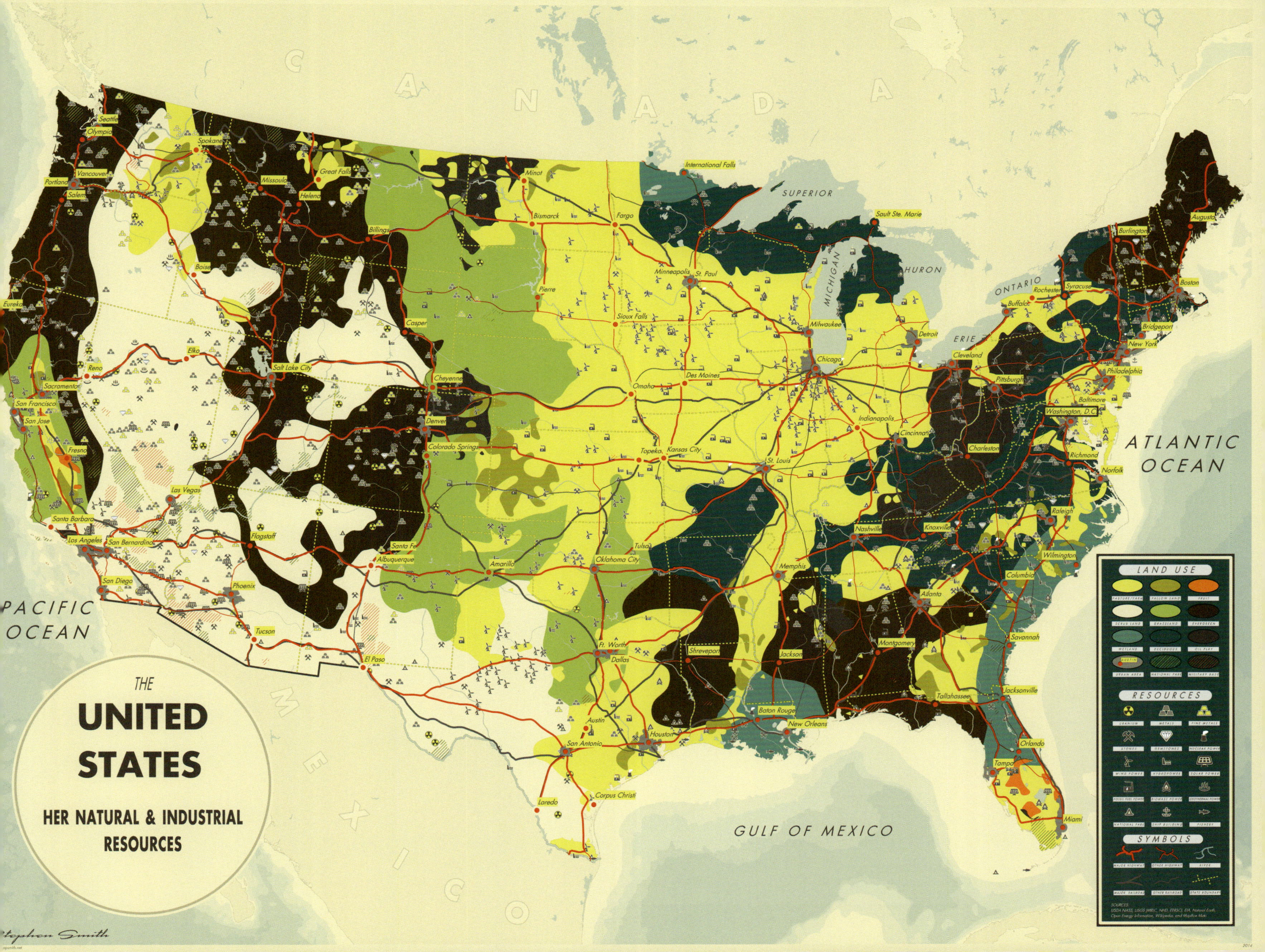
THE
UNITED
STATES
HER NATURAL & INDUSTRIAL
RESOURCES
CANADA
MEXICO
PACIFIC OCEAN
ATLANTIC OCEAN
GULF OF MEXICO
SUPERIOR
MICHIGAN
HURON
ONTARIO
ERIE
LAND USE
RESOURCES
SYMBOLS
Seattle
Olympia
Spokane
Vancouver
Portland
Salem
Missoula
Helena
Great Falls
Minot
International Falls
Bismarck
Fargo
Sault Ste. Marie
Augusta
Burlington
Billings
Boise
Eureka
Minneapolis
St. Paul
Pierre
Sioux Falls
Milwaukee
Rochester
Syracuse
Boston
Buffalo
Bridgeport
New York
Casper
Detroit
Cleveland
Chicago
Elko
Reno
Sacramento
Salt Lake City
Cheyenne
Omaha
Des Moines
Pittsburgh
Philadelphia
Baltimore
Washington, D.C.
San Francisco
San Jose
Denver
Colorado Springs
Indianapolis
Cincinnati
Charleston
Topeka
Kansas City
St. Louis
Richmond
Norfolk
Fresno
Las Vegas
Santa Barbara
Los Angeles
San Bernardino
Flagstaff
Santa Fe
Albuquerque
Tulsa
Nashville
Knoxville
Raleigh
Amarillo
Oklahoma City
Memphis
Wilmington
Columbia
San Diego
Phoenix
Atlanta
Tucson
Ft. Worth
Dallas
Shreveport
Jackson
Montgomery
Savannah
El Paso
Tallahassee
Jacksonville
Baton Rouge
New Orleans
Austin
San Antonio
Houston
Orlando
Tampa
Laredo
Corpus Christi
Miami
Stephen Smith

THOSE WHO DID NOT CROSS 2005 -2015

(December 3, 2011) A 29-year-old pregnant woman died from drinking seawater after the boat she embarked on in Libya went adrift in the Mediterranean Sea for 16 days. From 2005 to 2015, over 16,000 other individuals were reported dead or missing as they tried to reach European shores fleeing conflict and instability in Africa and the Middle East.

Libya has become a popular starting point for many journeys, with people traffickers exploiting the country's power vacuum and increasing lawlessness. The relatively short distance to Lampedusa encourages more people to risk the journey.

Each ● on this map shows where a person went missing or died on the Western, Central, and Eastern Mediterranean routes to Europe.

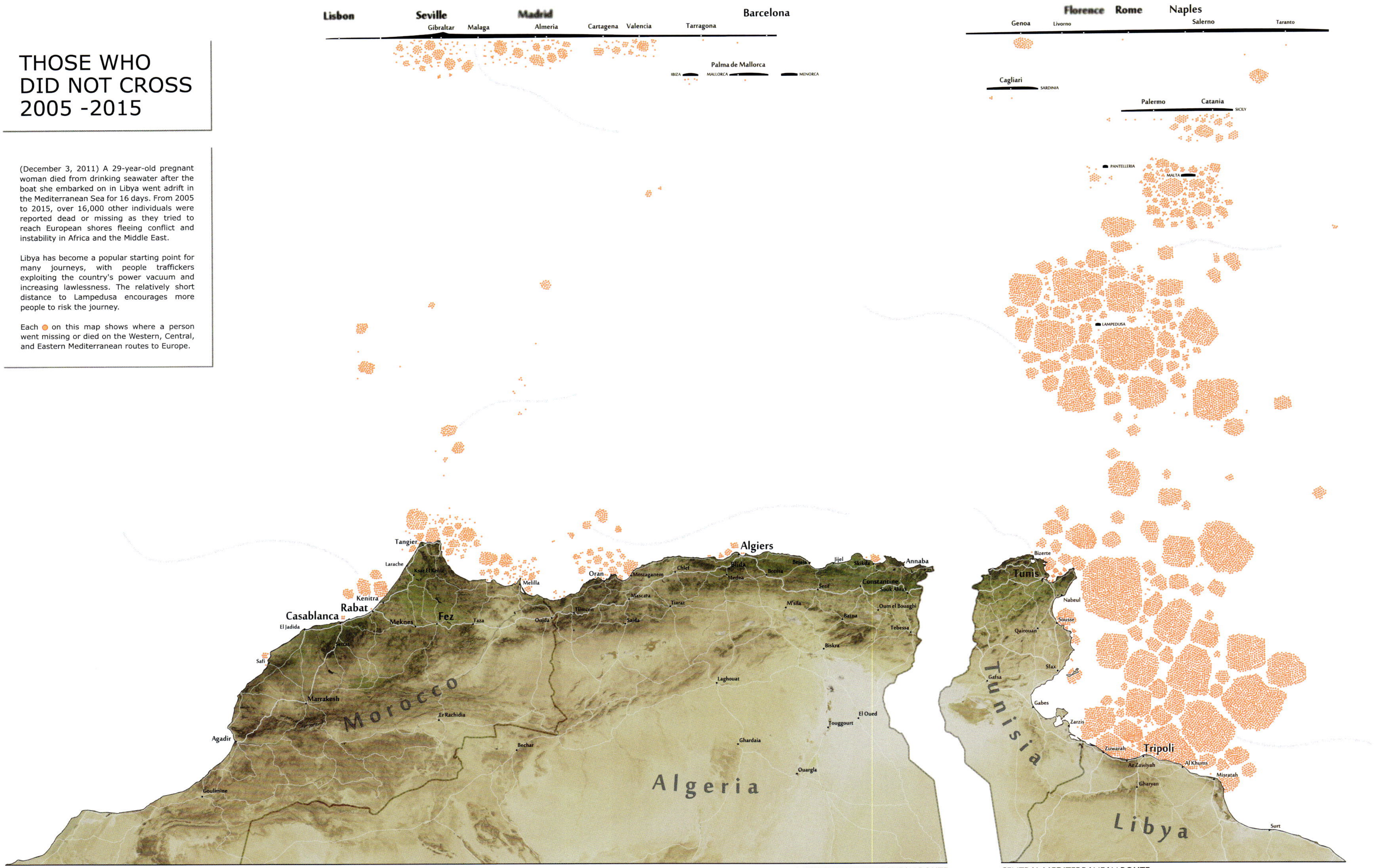

Levi Westerveld (2016). *Those Who Did Not Cross, 2005 - 2015*. 1:3,000,000. Lambert Conformal Conic projection, standard parallels at 30°N and 40°N.
Sources: The Migrant Files, USGS, Natural Earth.

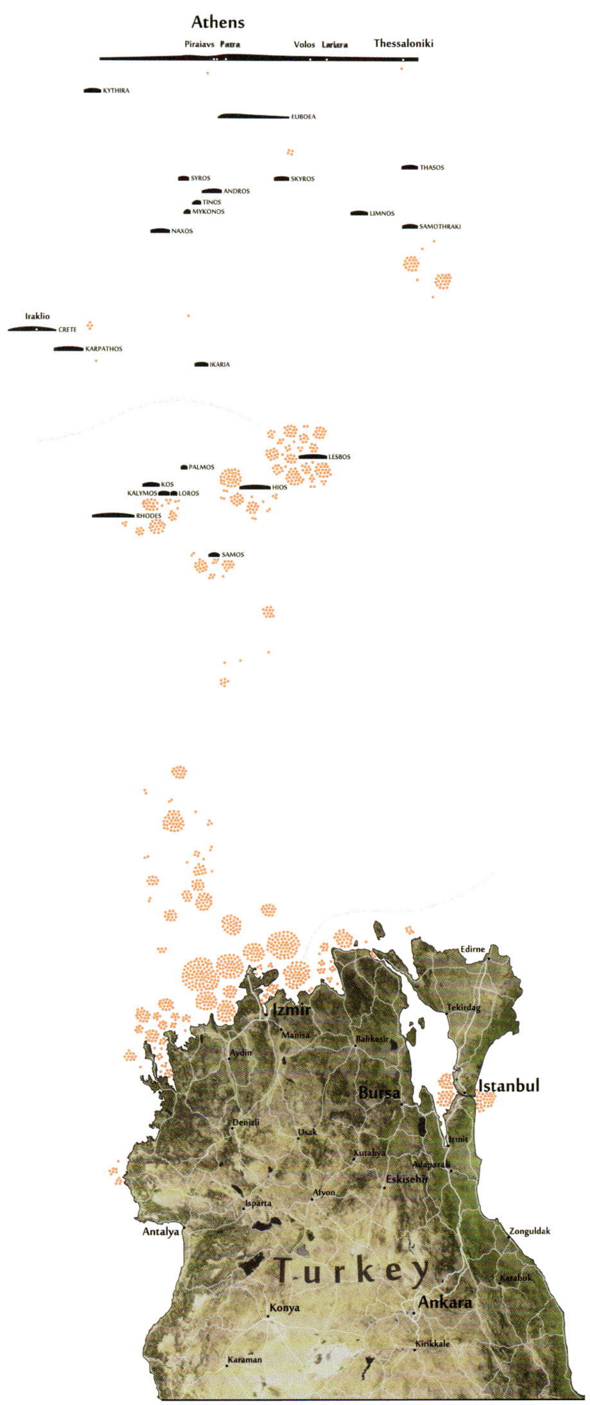

EASTERN MEDITERRANEAN ROUTE

THIS LAST MAP WAS ROTATED 90° WEST TO BETTER SHOW THE EASTERN MIGRATION ROUTE FROM TURKEY TO GREECE.

« Those Who Did Not Cross

Levi Westerveld

This map shows in a single image the people who lost their lives as they tried to reach European shores from 2005 to 2015. These men, women, and children were fleeing conflict and instability in Africa and the Middle East. My artistic decisions were driven by the question, "How can a map show the sheer number of victims in such a tragedy without abstracting the uniqueness of each individual?"

Assigning a unique dot to each victim helped to portray the unsettling number of recorded losses. I placed red dot symbols on the map, one by one, based on the location information of each accident. The lack of a distinct geometry behind the configuration of these islands of victims humanizes them and highlights their position suspended between their places of origin and planned destinations.

By flattening destination countries into black coastlines, the map illustrates what victims were waiting to see on the horizon. Distance and location are far from exact, but a wider portrayal of the Mediterranean Sea is closer to victims' experience of the journey, as they navigated in unseaworthy vessels, often using only handheld compasses.

Individual lines of blue text waving away from some of the clusters of victims relate the details of who died, how, and where they were going. And they leave us wondering: What about the stories behind all the other dots?

The North American Continent »

Anton Thomas

Our planet is far more miraculous than any fantasy world. From tundra to tropics, Earth is overflowing with incredible complexity. One need not draw dragons roaming the wilderness when Earth's fauna are so bizarre and beautiful. My map, "The North American Continent," is a tribute to reality. By vividly illustrating geographic data, I hope to bring attention to the dynamism and mystique of our world. Even on a continent tamed by the interstate and the Internet, a grand frontier is alive and well.

"The North American Continent" is an ongoing project—a pictorial map drawn completely by hand with coloured pencil and pen. It is a 59 x 47-inch cylindrical projection of the entire region, spanning Alaska to Panama and Greenland to the Caribbean. Currently around 80 percent complete, the map has thousands of features, including 330 animals and almost 500 cities and towns. From Los Angeles, California to Kugluktuk, Nunavut, each locale includes real skylines and landmarks.

The map began as a love letter to the continent. After several years of wandering the United States and Canada, I was stunned by the titanic scale and diversity on display. I hail from the remote archipelago of New Zealand, so the vast masterpiece of North American geography moved me. In an impatient world of 140-character attention spans, I wondered if an exceptionally detailed pictorial map could capture an audience's imagination. Perhaps it could even recast people's expectations about cartography.

North America offers a rich bounty of content. From ice caps to high desert, volcanoes to boundless prairie, the landscape is cast in high relief here. Our own imprint is just as charismatic: Soaring American skylines give way to constellations of hamlets high above the Arctic Circle. Magnificent Mayan ruins bejewel the same jungles that host Cancun's high-rise hotels.

As modern consumers flit through unlimited content in a heartbeat, it is worth remembering we are captivated most by what entertains us. By presenting North America in a compelling, creative way, I hope to make cartography captivating for everyone. If even the least enthusiastic observers can find themselves engrossed in the rich, endless detail, it may plant the seeds of geographic curiosity. For inspiration we need only look around. Adventure is everywhere in this world, and maps help us find it.

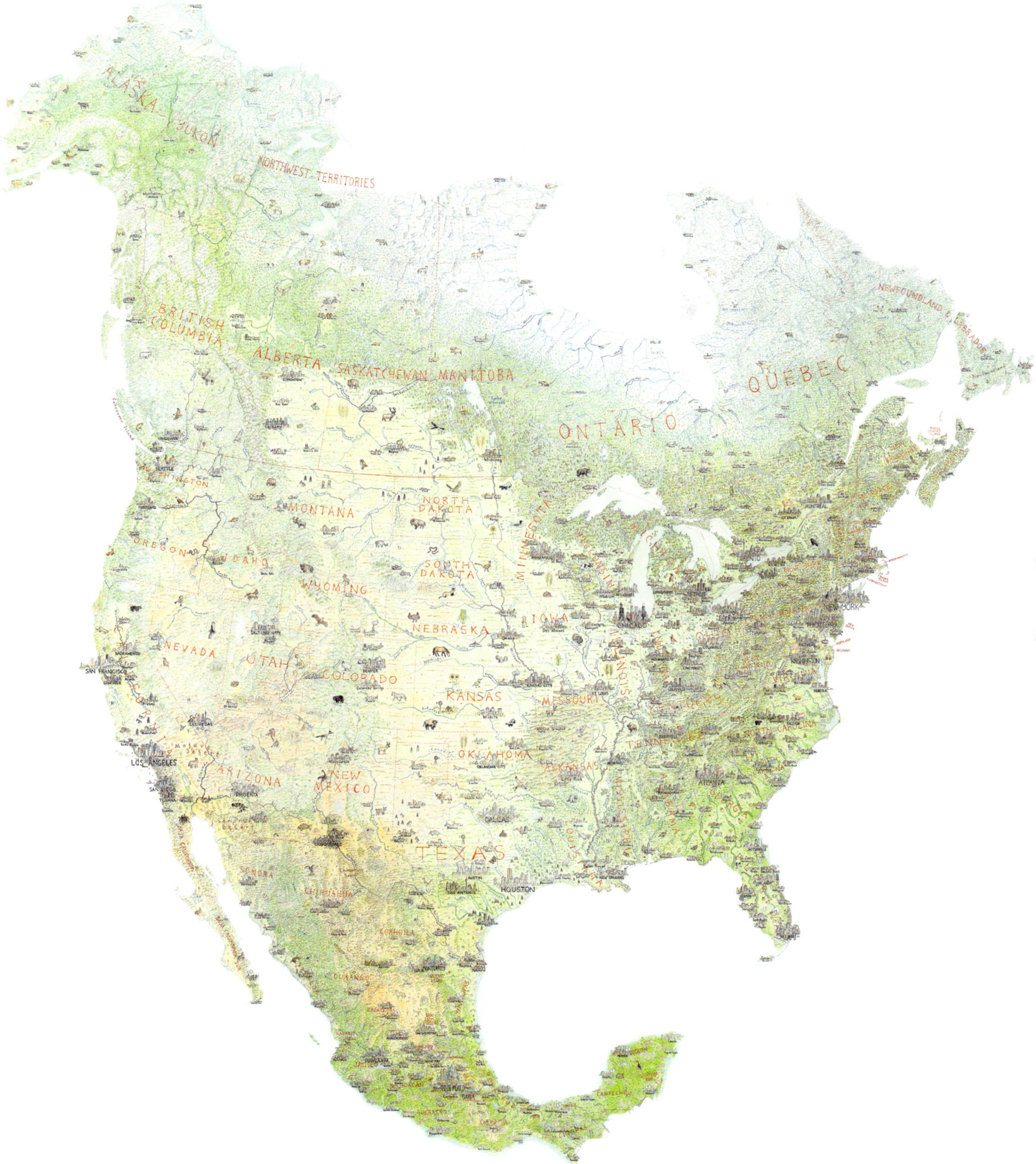

...on a continent tamed by the interstate and the Internet, a grand frontier is alive and well.

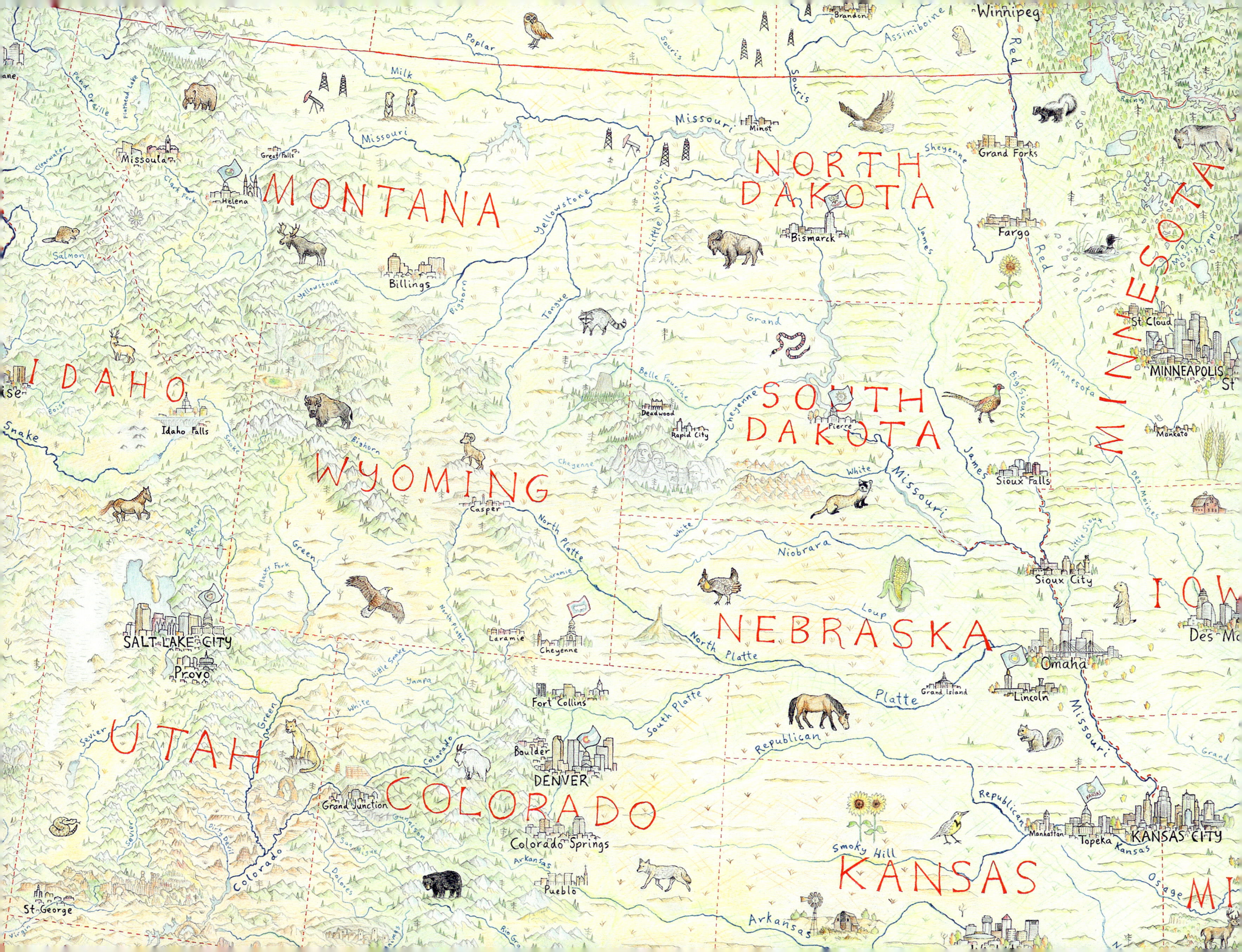
Winnipeg
Brandon
Assiniboine
Red
Poplar
Souris
Milk
Missouri
Minot
Pend Oreille
Flathead Lake
Missoula
Great Falls
Helena
Clark Fork
Clearwater
Salmon
MONTANA
NORTH DAKOTA
Grand Forks
Sheyenne
Bismarck
Fargo
James
Rainy
Mississippi
Yellowstone
Little Missouri
Billings
Bighorn
Tongue
Grand
Belle Fourche
Deadwood
Rapid City
Cheyenne
SOUTH DAKOTA
Pierre
White
Sioux Falls
Big Sioux
Minnesota
St Cloud
MINNEAPOLIS
Mankato
MINNESOTA
Des Moines
IDAHO
Idaho Falls
Boise
Snake
WYOMING
Casper
North Platte
Niobrara
Little Sioux
Sioux City
Bear
Green
Black Fork
Laramie
Cheyenne
SALT LAKE CITY
Provo
NEBRASKA
Loup
Omaha
Lincoln
Grand Island
Platte
Little Snake
Yampa
South Platte
Fort Collins
Boulder
DENVER
Republican
UTAH
Sevier
COLORADO
Grand Junction
Colorado
Gunnison
Colorado Springs
Arkansas
Pueblo
Dolores
St George
Virgin
Smoky Hill
KANSAS
Manhattan
Topeka
Kansas
KANSAS CITY
Osage

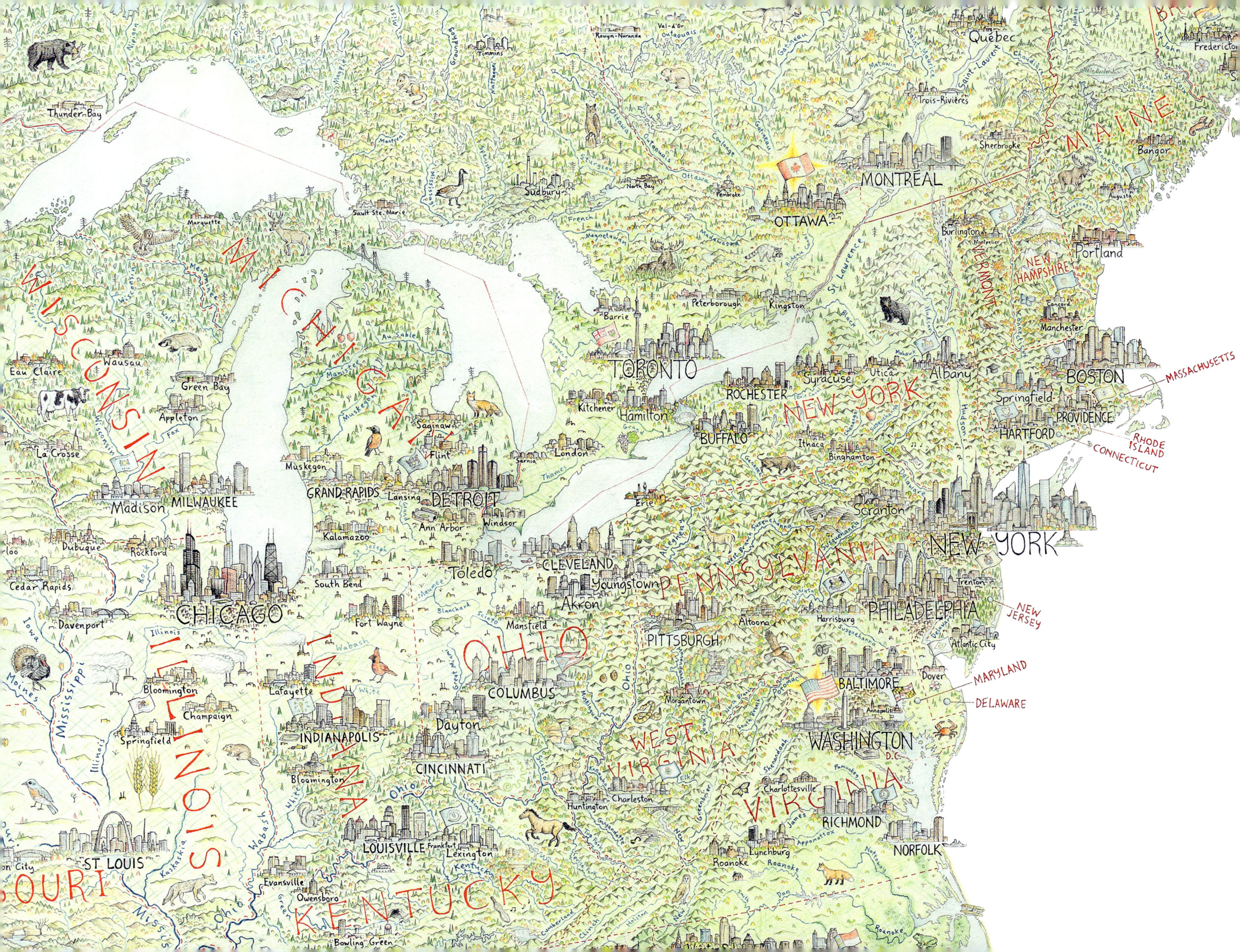

Thunder Bay
Timmins
Rouyn-Noranda
Val-d'Or
Québec
Fredericton
Trois-Rivières
Sherbrooke
MAINE
Bangor
Augusta
MONTREAL
OTTAWA
Sudbury
North Bay
Pembroke
Sault Ste. Marie
Marquette
MICHIGAN
WISCONSIN
Burlington
Montpelier
VERMONT
NEW HAMPSHIRE
Portland
Concord
Manchester
Peterborough
Kingston
Barrie
TORONTO
Hamilton
Kitchener
London
Sarnia
Wausau
Eau Claire
Green Bay
Appleton
La Crosse
Madison
MILWAUKEE
Muskegon
GRAND RAPIDS
Lansing
Saginaw
Flint
DETROIT
Windsor
Ann Arbor
Kalamazoo
ROCHESTER
Syracuse
Utica
Albany
BUFFALO
NEW YORK
Ithaca
Binghamton
Springfield
BOSTON
MASSACHUSETTS
PROVIDENCE
HARTFORD
RHODE ISLAND
CONNECTICUT
Erie
Scranton
NEW YORK
Trenton
Dubuque
Rockford
Cedar Rapids
Davenport
CHICAGO
South Bend
Toledo
CLEVELAND
Youngstown
Akron
PENNSYLVANIA
PHILADELPHIA
NEW JERSEY
Fort Wayne
Mansfield
OHIO
PITTSBURGH
Altoona
Harrisburg
Atlantic City
ILLINOIS
INDIANA
Lafayette
Bloomington
Champaign
Springfield
COLUMBUS
Morgantown
BALTIMORE
Dover
MARYLAND
Annapolis
DELAWARE
INDIANAPOLIS
Dayton
WASHINGTON
D.C.
WEST VIRGINIA
CINCINNATI
Bloomington
Charlottesville
VIRGINIA
Huntington
Charleston
RICHMOND
LOUISVILLE
Frankfort
Lexington
Lynchburg
NORFOLK
ST LOUIS
Roanoke
Evansville
Owensboro
KENTUCKY
MISSOURI
Bowling Green
Mississippi
Ohio
Wabash
Illinois
St Lawrence
Hudson
Ottawa

UFO SIGHTINGS

SELF-REPORTED SIGHTINGS SINCE 1925

JOHN NELSON
IDV SOLUTIONS

SIGHTINGS DATA | NATIONAL UFO REPORTING CENTER
2014 POPULATION | US CENSUS BUREAU

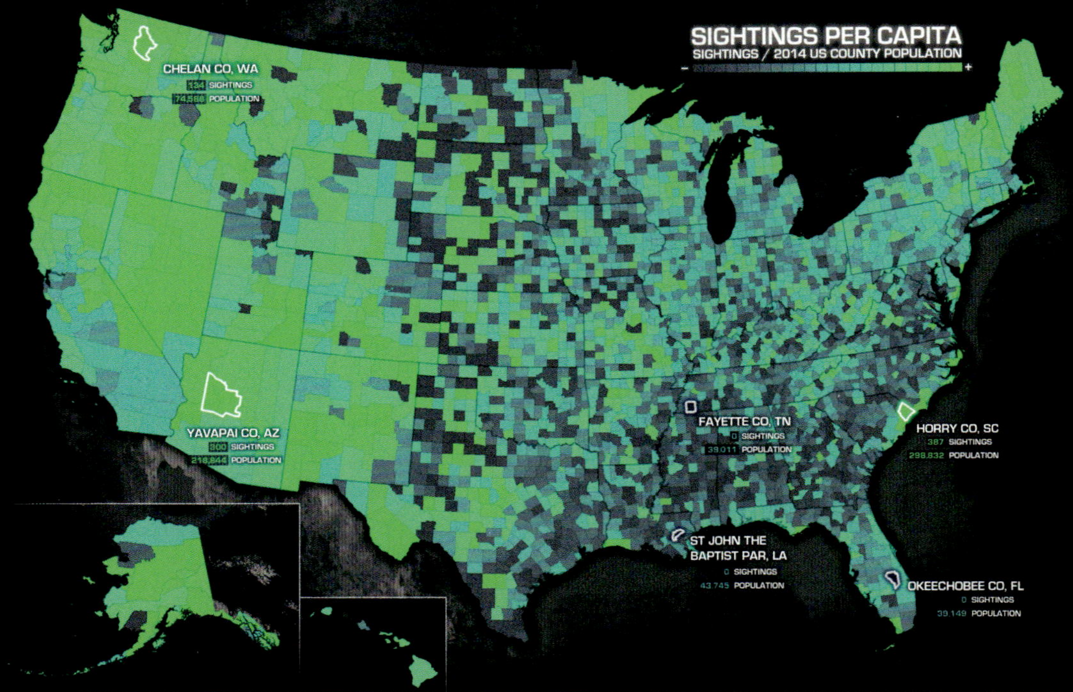

SHAPE TRENDING

AS A PROPORTION OF OTHER REPORTED SHAPES THAT YEAR. TREND LINES BEGIN IN 1964.

PREVALENCE

SYMBOLS REPRESENT 100 SIGHTINGS

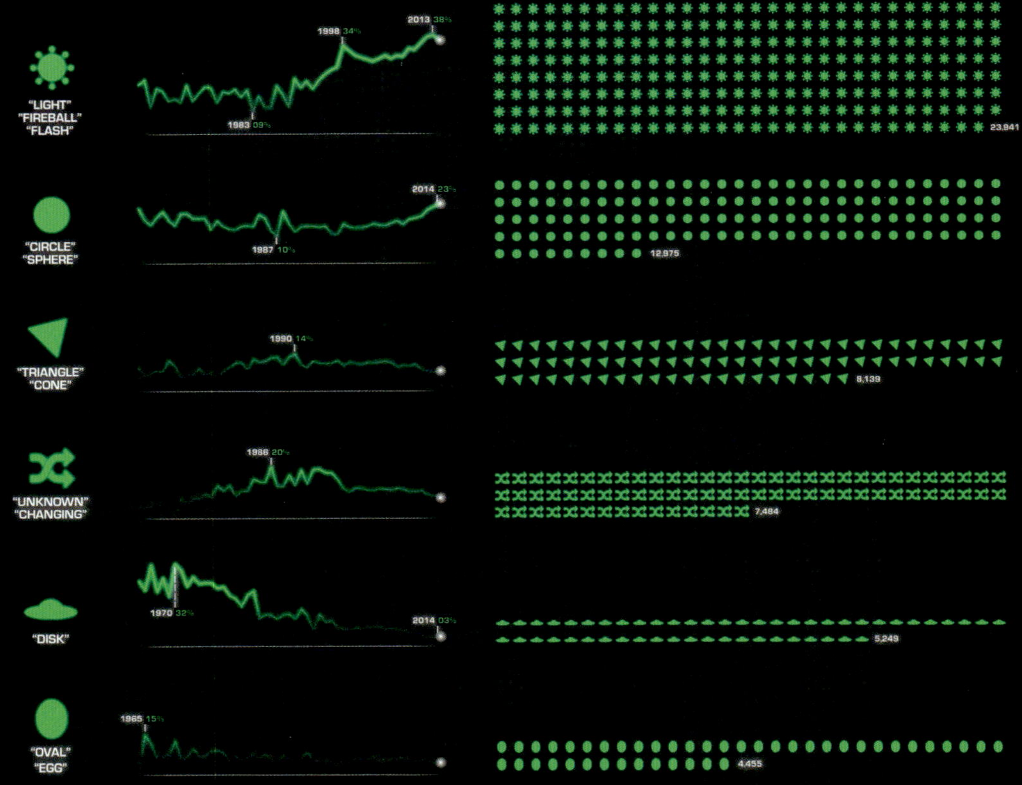

16,351 LIGHT	6,258 FIREBALL	5,249 DISK	2,409 FORMATION	1,332 FLASH	1,154 DIAMOND	721 TEARDROP
7,828 TRIANGLE	5,778 UNKNOWN	5,245 SPHERE	1,994 CIGAR	1,296 RECTANGLE	940 CHEVRON	311 CONE
7,730 CIRCLE	5,524 OTHER	3,723 OVAL	1,960 CHANGING	1,261 CYLINDER	732 EGG	[illegible] CROSS

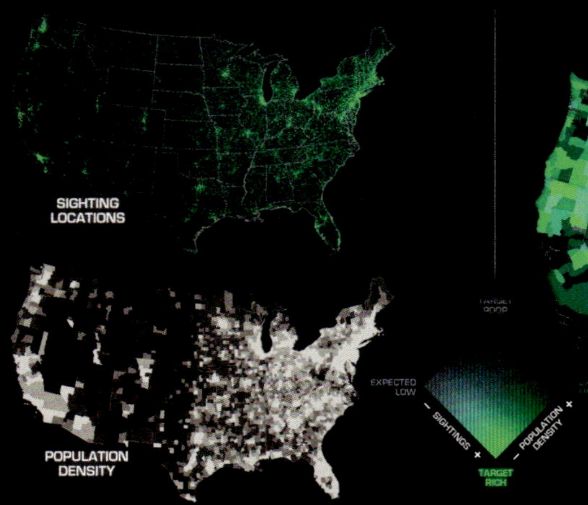

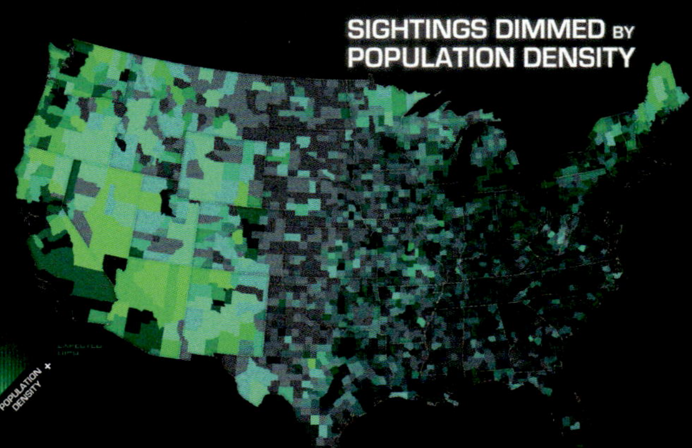

SIGHTING TIME, SEASONALLY

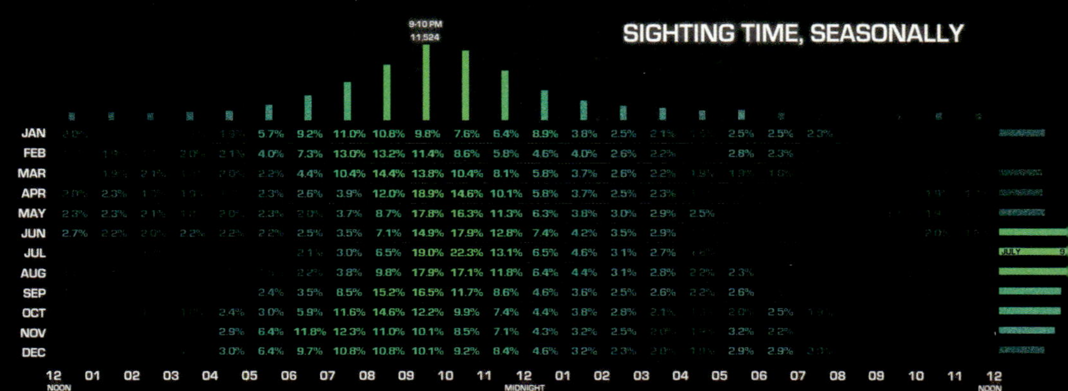

Crowd-sourced data tends to reflect how the crowd sees itself, and even the most alien topics become little portraits of ourselves.

« UFO Sightings

John Nelson

This map descended from the glittering night sky with a suddenness and jolt of energy that I find difficult to describe to this day. But I'll try. . . .

Spanning almost 100 years of self-reported UFO sightings, this map plots the geographic terrain of the events, seasonal trends, and how witnesses describe the UFOs' shapes. Rendered in alien green over a dark setting, the elements have a noisy CRT monitor vibe that mimics bogie trails you might see over the shoulder of a CIA analyst.

I took care to account for underlying population so that the result was not just another population map dressed up as something else. The sightings-per-capita map shows a normalized impression of where reported sightings were more likely to happen. And since people may be more likely to notice something unidentifiable in a darker, clearer sky, a companion bi-variate map applies transparency to areas of high population density. This map is Mega-Normalized!

Data that spans decades is a gift to cartographers. Long-term social data, such as UFO sightings, exhibits so many trends and such structure. The fun is finding a vein with a good story and letting that reveal itself.

What I found most interesting in the data was our own meandering American social terrain. In the mid-twentieth century, we were certain that "disks" filled the skies, but in the 1980s and 1990s we used less concrete descriptions of the objects, like "unknown" and "changing." Recently, more sensational, but still vague, descriptions like "light," "fireball," and "flash" have quadrupled in popularity. Crowd-sourced data tends to reflect how the crowd sees itself, and even the most alien topics become little portraits of ourselves.

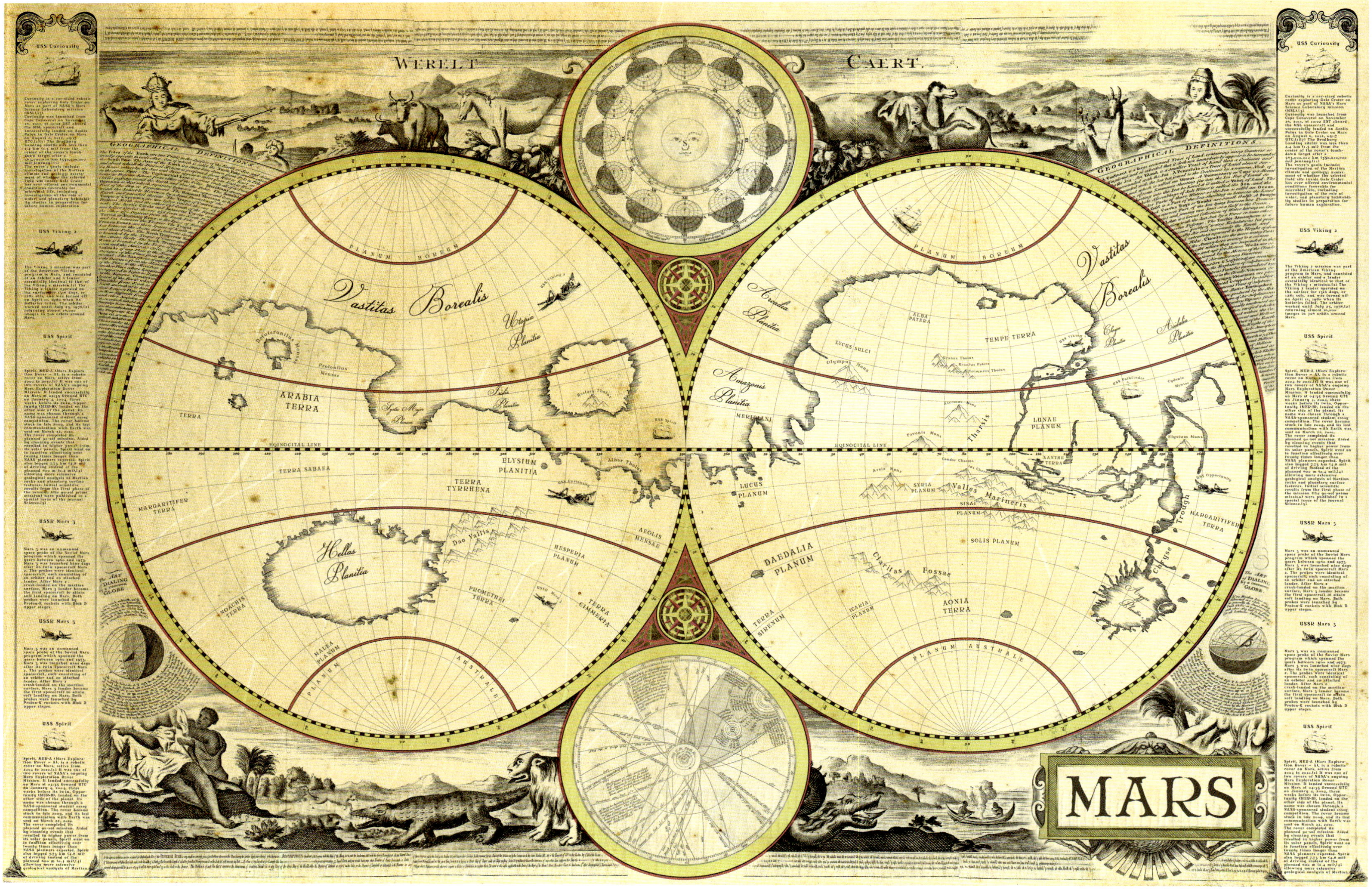
WERELT CAERT.
MARS
GEOGRAPHICAL DEFINITIONS
USS Curiosity
USS Viking 2
USS Spirit
USSR Mars 3
Vastitas Borealis
Utopia Planitia
PLANUM BOREUM
ARABIA TERRA
TERRA
EQUINOCTIAL LINE
TERRA SABAEA
TERRA TYRRHENA
ELYSIUM PLANITIA
MARGARITIFER TERRA
Hellas Planitia
NOACHIS TERRA
PROMETHEI TERRA
HESPERIA PLANUM
AEOLIS MENSAE
TERRA CIMMERIA
MALEA PLANUM
PLANUM AUSTRALE
Arcadia Planitia
Amazonis Planitia
ALBA PATERA
TEMPE TERRA
LYCUS SULCI
Olympus Mons
Tharsis
LUNAE PLANUM
XANTHE TERRA
LUCUS PLANUM
SYRIA PLANUM
Valles Marineris
SOLIS PLANUM
DAEDALIA PLANUM
Claritas Fossae
ICARIA PLANUM
AONIA TERRA
TERRA SIRENUM
Chryse Planitia
Acidalia Planitia
MERIDIANI
Margaritifer Terra
The Art of Dialing Globe

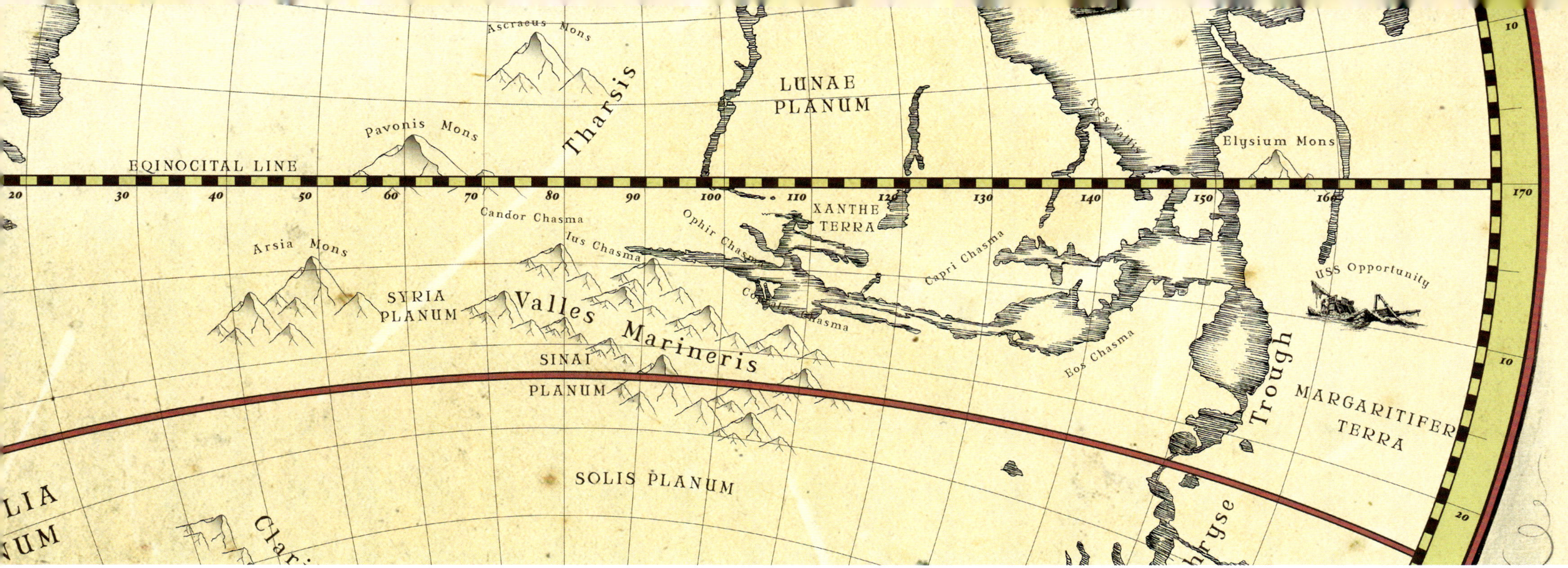

« Mars Cartography

Michael Allen Nesmith

Antique maps are known for their elaborate borders, strange sea monsters, and unique type hierarchies that use a mixture of calligraphy and serif typeface. These maps offer much more than simple navigational information: They tell stories about newly "discovered"—but often long-inhabited—places and offer insight into what a traveller might see there. Style and design elements that distinguish antique maps are often borrowed, adapted, and then lost over time.

Given that the landscape of Mars is both new to cartographers and very old, it seemed fitting to tell the planet's story in the style of an antique New World map. I began by tracing a large topographic shaded relief map of Mars produced by the Mars Orbiter Laser Altimeter (MOLA), an important instrument aboard the Mars Global Surveyor (MGS). Looking at MOLA's color-coded map, I used the yellow "sea level" shade as a guide to determine a distinct "shoreline." I was fascinated to note what kind of landmasses, seas, rivers, and oceans Mars has in this context.

I drafted colonial ships (active missions on Mars), shipwrecks (defunct and/or failed missions), and wagons (rovers) separately, then transplanted them to the main map at their current locations. A fictional but scientifically sound story about Mars emerged.

This pet project helped to illuminate the peculiar struggles of doing cartography by hand. For example, when I was about halfway through the entire process, I paused before moving on to another secondary master layer only to find out that I had been using the wrong point size for my pen, meaning I had to scrap the whole thing to start anew. I have developed a deep sense of empathy for early cartographers and hope to pursue other uncharted territories in the future.

Dolomiti di Brenta Hiking Trails Map »

Marco Barbieri

When I was a boy vacationing in the Alps, I used to buy hiking maps—mostly Swiss and Italian topographic maps, at 1:25,000 scale. I started collecting these beautiful maps and never stopped. I collected them not only to find my way in the mountains, but also because they fascinated me aesthetically, with their rich detail, careful choice of typefaces, the placement of each name in the right position, and the meticulous drawing of rocks and hillshade. Careful design and the hand of each individual cartographer were apparent in the works.

This map of the Brenta Dolomites is strongly inspired by classic topographic hiking maps, with the addition of some stylistic innovations. My general aim is to produce a less technical and more natural representation than in the maps of the past. I'm able to achieve this through the use of digital data and techniques and careful choice of colors.

Three different shades of green represent the forests, shrubland, and grassland. The pinkish gray and light blue colors indicate rocky areas and glaciers. Hillshade and contour lines are derived from a detailed digital terrain model and depict the geomorphology of the area in high definition and high fidelity. In general, the colors of the map and the hillshade are intentionally soft, to make place names and red signs related to hiking clearly legible. In addition, I pay close attention to the fact that the map is published in two versions: one for printing and another for web publishing.

The overall result is a map in which it is easy to read the morphology and nature of the land. The viewer can identify large inhabited valleys, wooded slopes, high mountain pastures, barren rocky plateaus, steep walls, and the great glaciers and peaks of the Brenta Dolomites. Viewers can also see the names of all of these geographic features. The map becomes a precise and clear navigation instrument for hikers and mountaineers who want to explore the *Dolomiti di Brenta*.

The web version of the map is available at http://brenta.webmapp.it. Data sources include OpenStreetMap contributors, under ODbL; Società degli Alpinisti Tridentini, under ODbL; and Provincia di Trento, under ODbL© Provincia di Trento, under ODbL

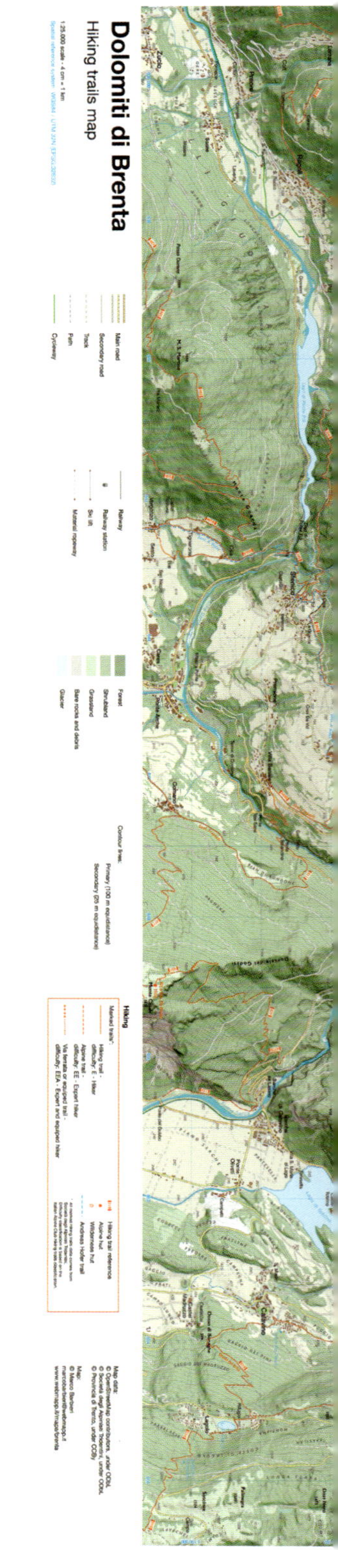

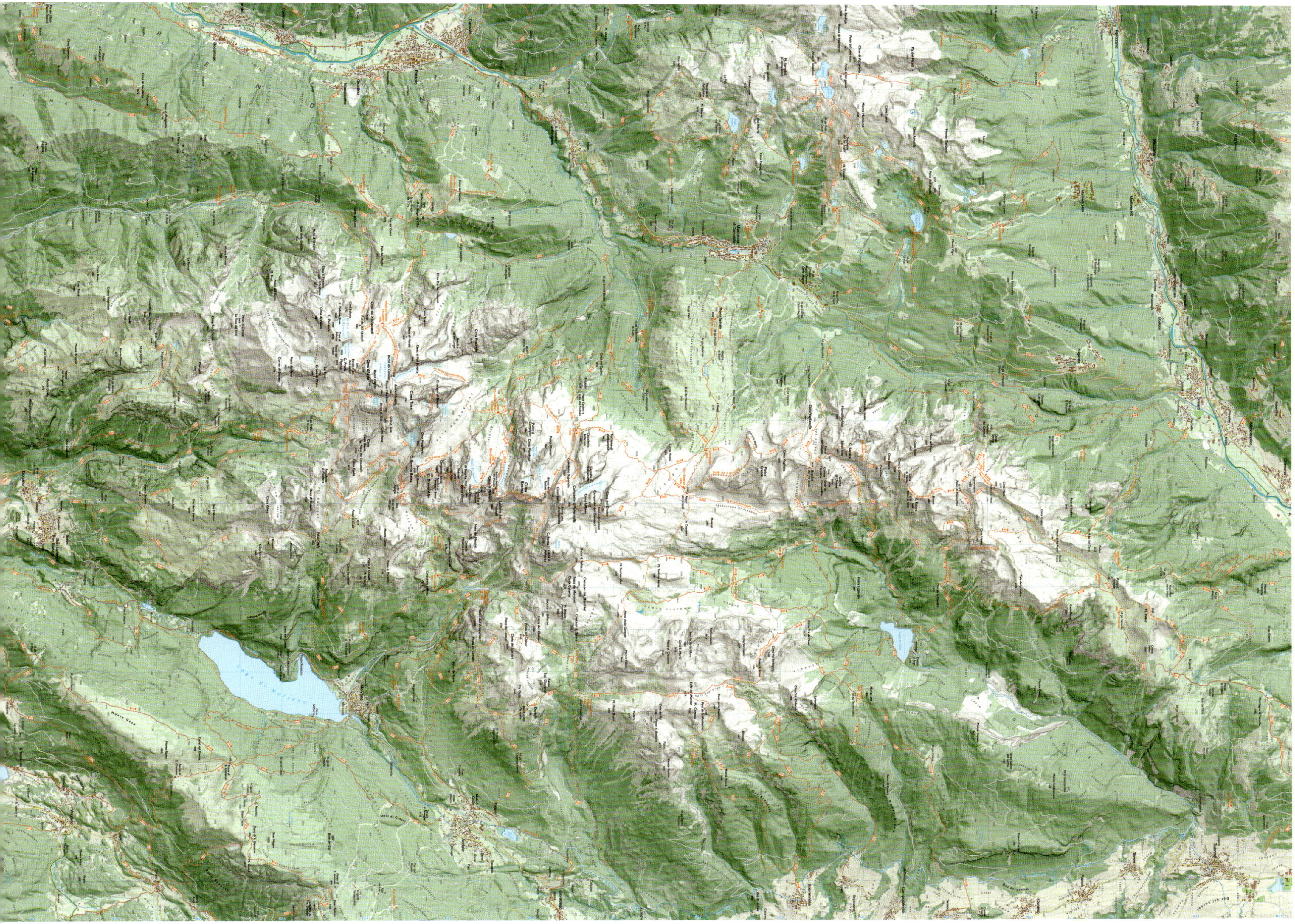

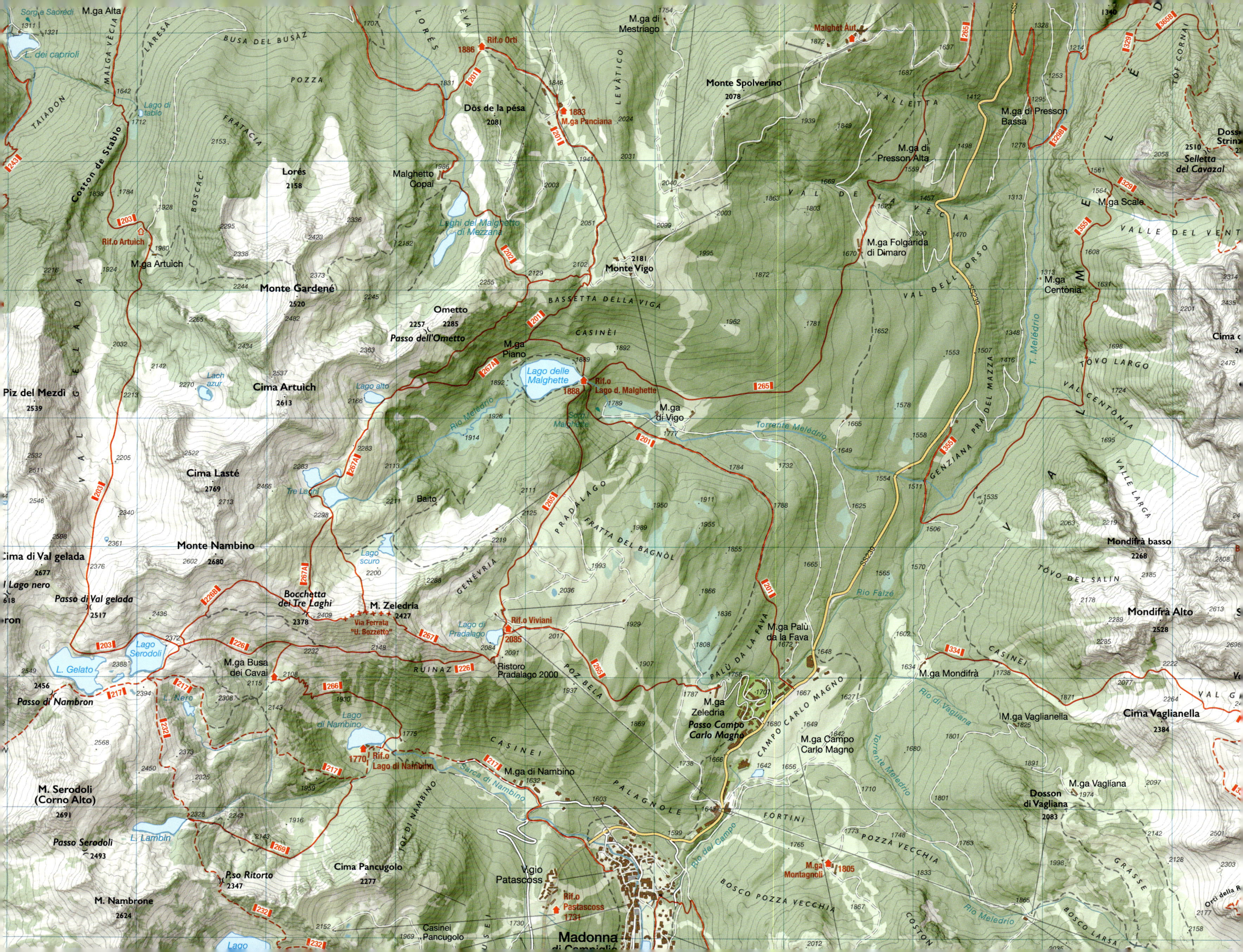

M.ga Alta
L. dei caprioli
Rif.o Orti
1886
Rif.o Artuich
M.ga Artuich
Lago di Tablo
Coston de Stablo
Dòs de la pèsa
2081
M.ga Panciana
1883
M.ga di Mestriago
Malghet Aut
Monte Spolverino
2078
Lorés
2158
Malghetto Copai
Laghi del Malghetto di Mezzana
Monte Vigo
2181
Monte Gardené
2520
Ometto
2285
Passo dell'Ometto
M.ga Piano
Lago delle Malghette
Rif.o Lago d. Malghette
1888
M.ga di Vigo
Bassetta della Viga
Casinèi
Val de la Véla
Val dell'Orso
M.ga Folgarida di Dimaro
M.ga di Presson Alta
M.ga di Presson Bassa
M.ga Scale
M.ga Centonia
Val Centonia
Tòvo Largo
Selletta del Cavazal
Piz del Mezdì
2539
Cima Artuich
2613
Lago alto
Torrente Meledrio
Cima Lasté
2769
Tre Laghi
Baito
Pradalago
Fratta del Bagnòl
Genziana Pra del Mazza
Valle Larga
Mondifrà basso
2268
Tòvo del Salin
Mondifrà Alto
2528
Monte Nambino
2680
Cima di Val gelada
2677
Lago nero
Passo di Val gelada
2517
Bocchetta dei Tre Laghi
2378
M. Zeledria
2427
Via Ferrata "U. Bozzetto"
Lago scuro
Lago di Pradalago
Rif.o Viviani
2085
Ristoro Pradalago 2000
Genèvria
Poz Bela
Palù da la Fava
M.ga Palù da la Fava
Rio Falzè
L. Gelato
Lago Serodoli
M.ga Busa dei Cavai
Ruinaz
Passo di Nambron
L. Nero
Lago di Nambino
Rif.o Lago di Nambino
1770
M.ga Zeledria
Passo Campo Carlo Magno
Campo Carlo Magno
M.ga Campo Carlo Magno
M.ga Mondifrà
Casinei
M.ga Vaglianella
Cima Vaglianella
2384
Rio di Vagliana
Torrente Meledrio
M.ga Vagliana
Dosson di Vagliana
2083
M.ga di Nambino
Sarca di Nambino
Palagnole
Fortini
Pozza Vecchia
Grasse
M. Serodoli (Corno Alto)
2691
Passo Serodoli
2493
L. Lambin
Cima Pancugolo
2277
V.gio Patascoss
Rif.o Pastascoss
1731
Pso Ritorto
2347
M. Nambrone
2624
Casinei Pancugolo
Madonna di Campiglio
M.ga Montagnoli
1805
Bosco Pozza Vecchia
Rio del Campo
Coston
Bosco Larsa
Rio Meledrio
Val Gelada

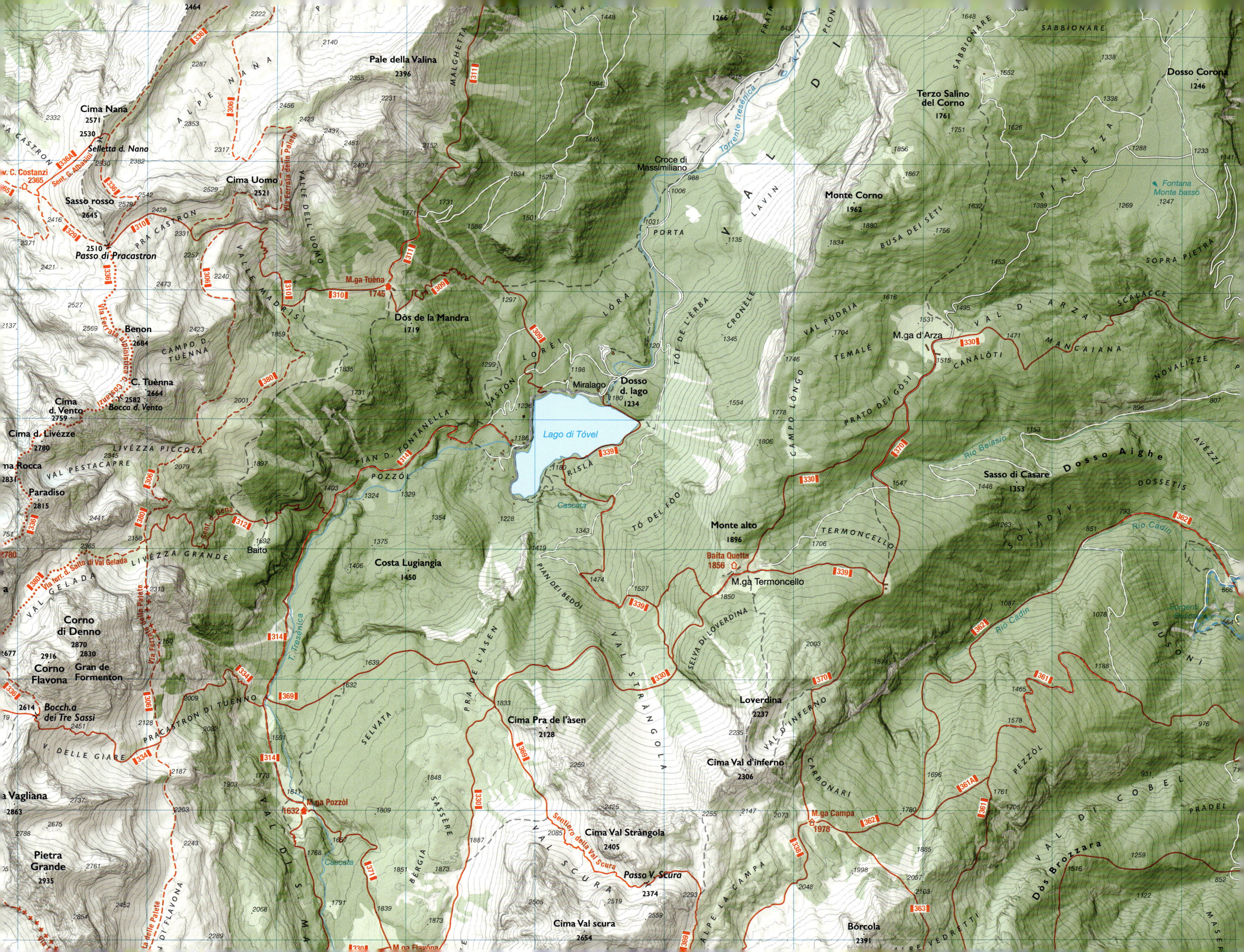

Pale della Valina
2396
Cima Nana
2571
Selletta d. Nana
Cima Uomo
2521
Sasso rosso
2645
Passo di Pracastron
2510
Pra Castron
Alpe Nana
Valle dell'Uomo
Valle Madris
Malghetta
M.ga Tuèna
1745
Dòs de la Mandra
1719
Benon
2684
Campo d. Tuenna
C. Tuènna
2664
Bocca d. Vento
2582
Cima d. Vento
2759
Cima d. Livèzze
2780
Livèzza Piccola
Val Pestacapre
Paradiso
2815
Pian d. Fontanella
Pozzòl
Baito
Livèzza Grande
Val Gelada
Corno di Denno
2870
Corno Flavona
2916
Gran de Formenton
2830
Bocch.a dei Tre Sassi
2614
V. delle Giare
Pracastron di Tuenno
Vagliana
2863
Pietra Grande
2935
Costa Lugiangia
1450
T. Tresènica
Selvata
Val di S. Ma
M.ga Pozzòl
1632
Cascata
M.ga Flavona
Bergia
Sassère
Pra de l'àsen
Cima Pra de l'àsen
2128
Pian dei Bedòi
Val Stràngola
Sentiero della Val Scura
Cima Val Stràngola
2405
Passo V. Scura
2374
Val Scura
Cima Val scura
2654
Lago di Tóvel
Miralago
Dosso d. lago
1234
Rislà
Castelàston
Lòra
Lorei
Porta
Croce di Massimiliano
Torrente Tresènica
Val di Plon
Lavin
Tóf de l'Èrba
Cronèle
Tó del Fóo
Monte alto
1896
Baita Quetta
1856
M.ga Termoncello
Termoncello
Selva di Loverdina
Loverdina
2237
Val d'Inferno
Cima Val d'inferno
2306
Carbonari
M.ga Campa
1978
Alpe la Campa
Bórcola
2391
Campo Longo
Val Pùdria
Temalé
Prato dei Gòsi
Busa dei Sèti
Monte Corno
1962
Terzo Salino del Corno
1761
Sabbionare
Pianezza
Dosso Corona
1246
Fontana Monte basso
Sopra Pietra
M.ga d'Arza
Val d'Arza
Canalòti
Mancaiana
Scalacce
Novalizze
Rio Belasio
Sasso di Casare
1353
Dosso Aighe
Dossetis
Soladiv
Rio Cadin
Avezzi
Busoni
Pezzòl
Val di Cobel
Pradel
Dòs Brozzara
Vedretti

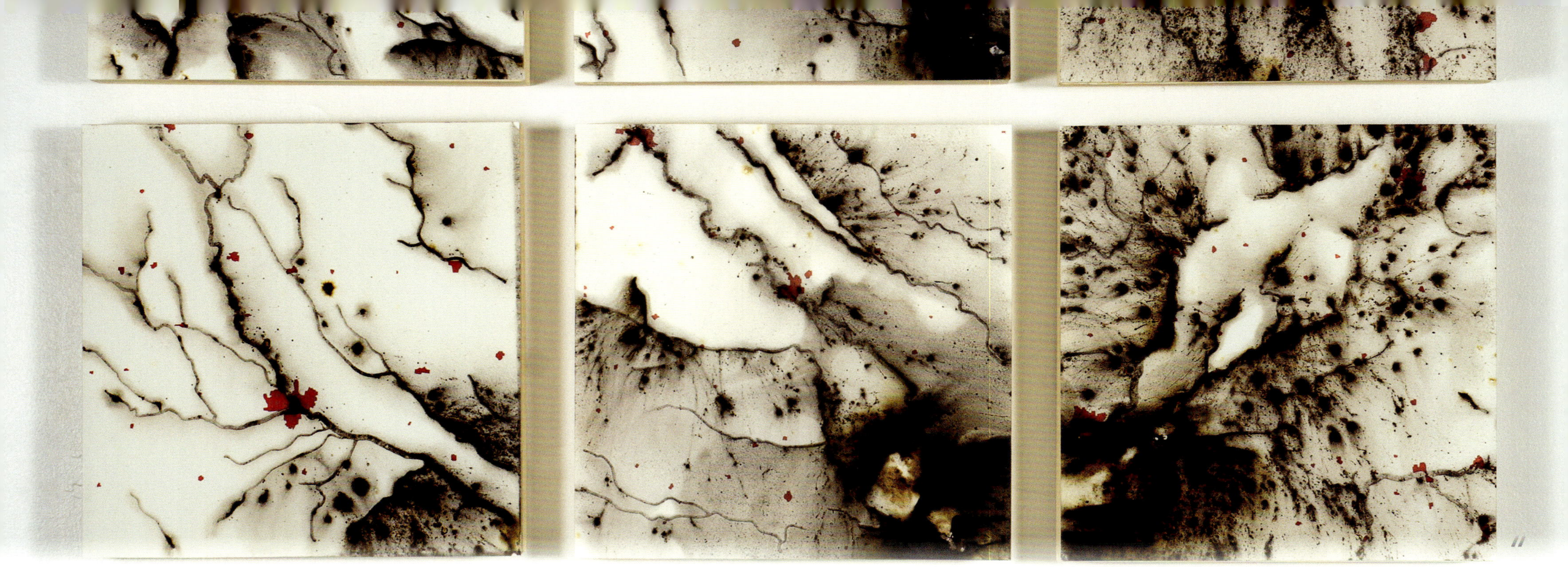

Mississippi River »

Matt Dooley, David Bergs, Matthew Bergs, Jason Blatz, Cynthia Brewer, Jeff Ferzoco, Donna Genzmer, Kyle Glowa, Sarah Harling, Chris Henrick, Diana Hoover, Brett Kallusky, Alison Link, Nick Martinelli, Ashley Nepp, Chelsea Nestel, Charlie Rader, Tim Stallmann, Diego Valadares, Steve Wolf, and Patrick Wood

"Mississippi River" is a collaborative artwork that grew out of the NACIS Gunpowder Mapping Workshop held in River Falls, Wisconsin, on October 17, 2015. The artwork is composed of fifteen 10 x 10-inch paper tiles, each showing a portion of the Mississippi River and adjacent drainage systems from Minneapolis (northwest) to Saint Louis (southeast). Each tile was assembled and ignited by a different workshop participant, resulting in 15 unique impressions of the region. The original tiles were then scanned, enlarged, printed, and mounted onto wooden frames for display.

The goal of the Gunpowder Mapping Workshop was to promote tangible interactions between makers and media and to foster connections among different groups of cartographers. Working in a new, unfamiliar medium with no expectations can help create a low-risk environment for creative collaboration. In a judgement-free space, makers feel comfortable to experiment, take risks, and even fail. Tangible interactions help people connect across the digital divide, promoting an egalitarian creative environment, where everyone has something to contribute.

Although novel in its application to mapping, gunpowder has been used as a medium by several contemporary artists. One notable example is Cai Guo-Qiang, a Chinese artist who has used gunpowder extensively in drawings and large-scale performance pieces. The technique used to create this piece closely follows a method taught by University of Wisconsin art professor Randy Johnston.

The resulting work, "Mississippi River," invites viewers to explore the visual beauty of river systems and to contemplate their relationship with the natural world. Whether the entire piece is seen as abstract art or a structured river map, each tile serves as a reminder that the finished piece is less important than the atmosphere created in the making of it.

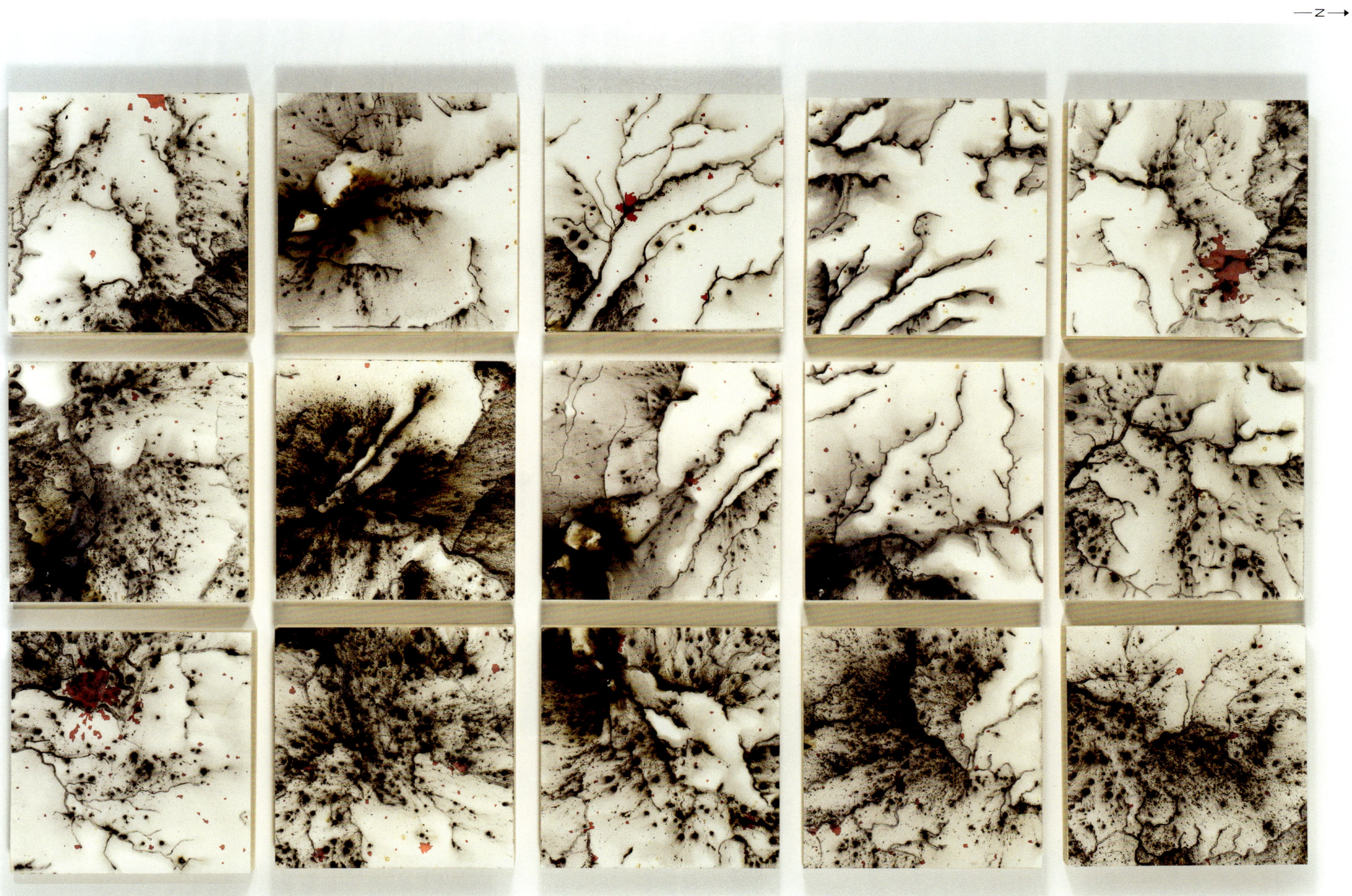

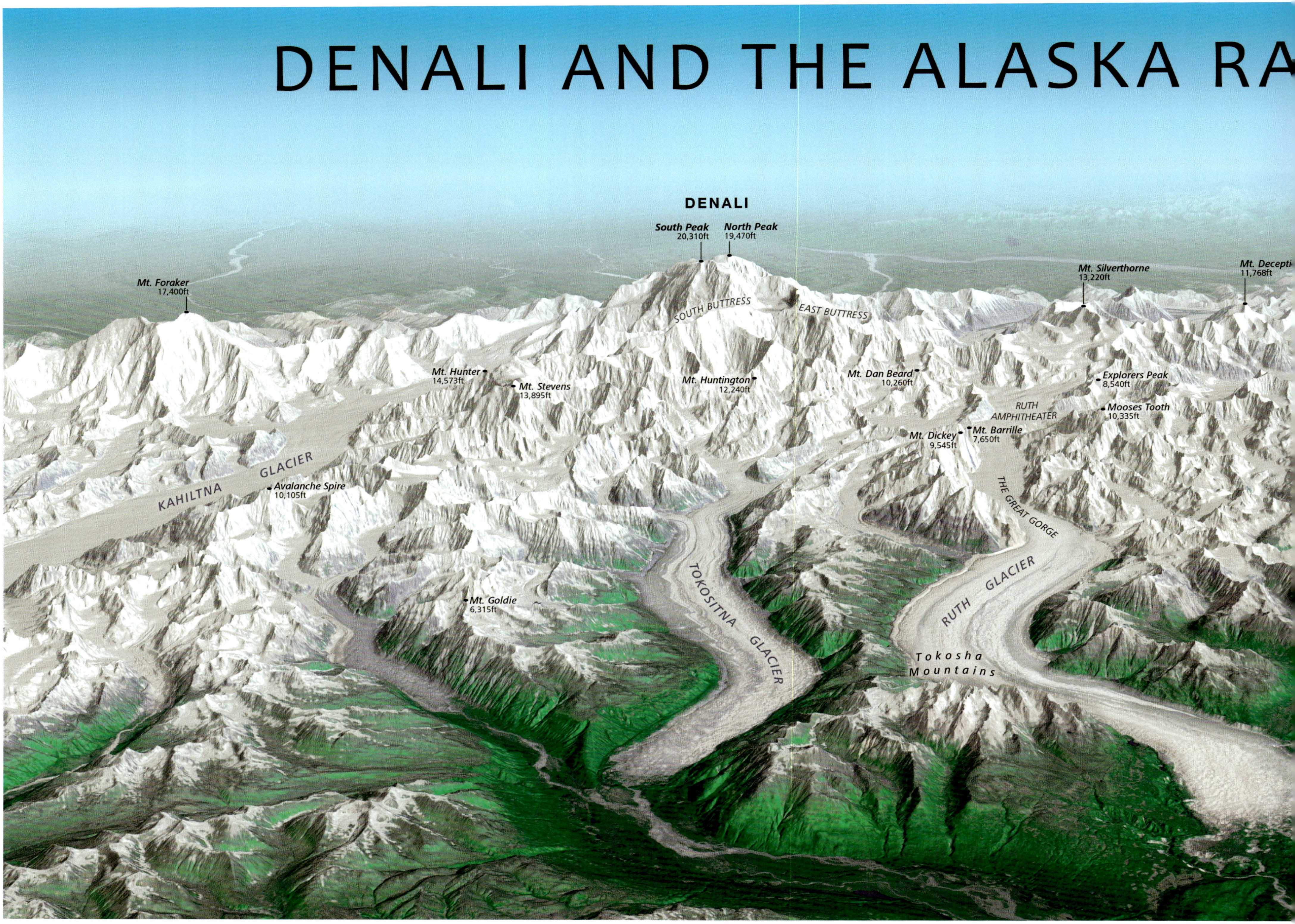
DENALI AND THE ALASKA RA
DENALI
South Peak
20,310ft
North Peak
19,470ft
Mt. Foraker
17,400ft
Mt. Silverthorne
13,220ft
Mt. Decepti
11,768ft
SOUTH BUTTRESS
EAST BUTTRESS
Mt. Hunter
14,573ft
Mt. Stevens
13,895ft
Mt. Huntington
12,240ft
Mt. Dan Beard
10,260ft
Explorers Peak
8,540ft
RUTH AMPHITHEATER
Mooses Tooth
10,335ft
Mt. Dickey
9,545ft
Mt. Barrille
7,650ft
KAHILTNA GLACIER
Avalanche Spire
10,105ft
THE GREAT GORGE
Mt. Goldie
6,315ft
TOKOSITNA GLACIER
RUTH GLACIER
Tokosha Mountains

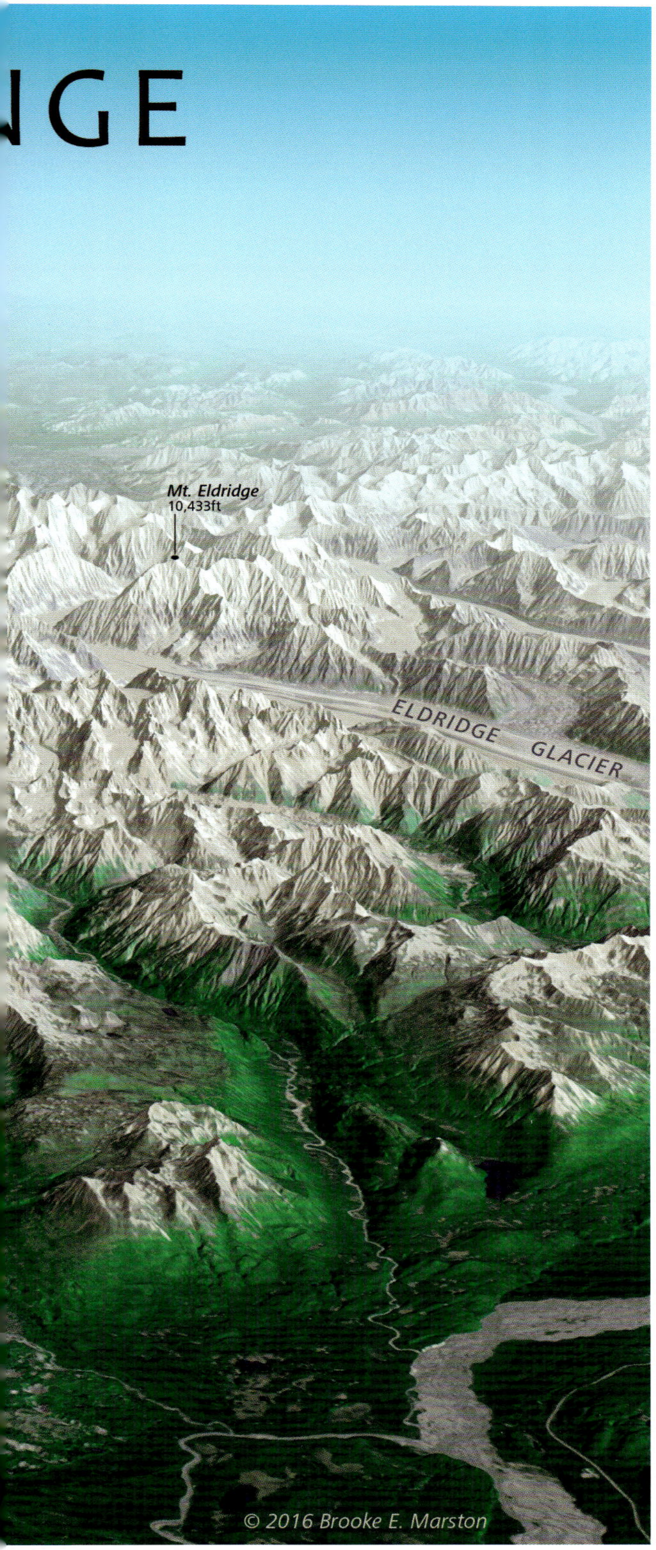

« Denali and the Alaska Range

Brooke Marston

This panoramic map of Denali and the Alaska Range was inspired by Austrian cartographer Heinrich Berann's work blending artistic ingenuity with natural realism. The painting of Denali, meaning "the great one," was Berann's final panorama, and since its publication in 1994, panoramic maps of Denali have been scarce. My own passion for relief representation and the renaming of North America's highest peak to its Alaska Native name motivated me to create this map.

I explored the advantages and limitations of perspective, orientation, and vertical exaggeration in three-dimensional terrain and incorporated aspects of Berann's distinct style to create a cartographic panorama of this mesmerizing place. The oblique perspective captures the grandness of Denali and the rugged expansiveness of the Alaska Range in one continuous view. The southeastern viewpoint highlights the long and narrow Alaska Range as it arcs to the northwest. The majestic Denali is featured in the center of the map with the range wrapping around and framing the peak. The Kahiltna, Tokositna, Ruth, and Eldridge Glaciers direct the viewer from lowlands in the foreground, up the span of the glaciers, to their nexus at the base of Denali.

Berann selectively rotated or rearranged mountains when painting panoramas by hand, but this map depicts a more realistic landscape with natural-color satellite imagery draped over the terrain. Berann's panorama is reminiscent of a landscape painting, using bright, saturated colors. To evoke a similar effect, I used bright green vegetation in the lowlands to suggest spring tones. High-elevation peaks were rendered in soft grays, whites, and blues recalling a wintery chill. The striking contrast between these color palettes invites the viewer to experience the landscape in a more visceral manner. The overall effect is to portray the scale and grandeur of this dramatic wilderness from the perspective of a soaring bald eagle.

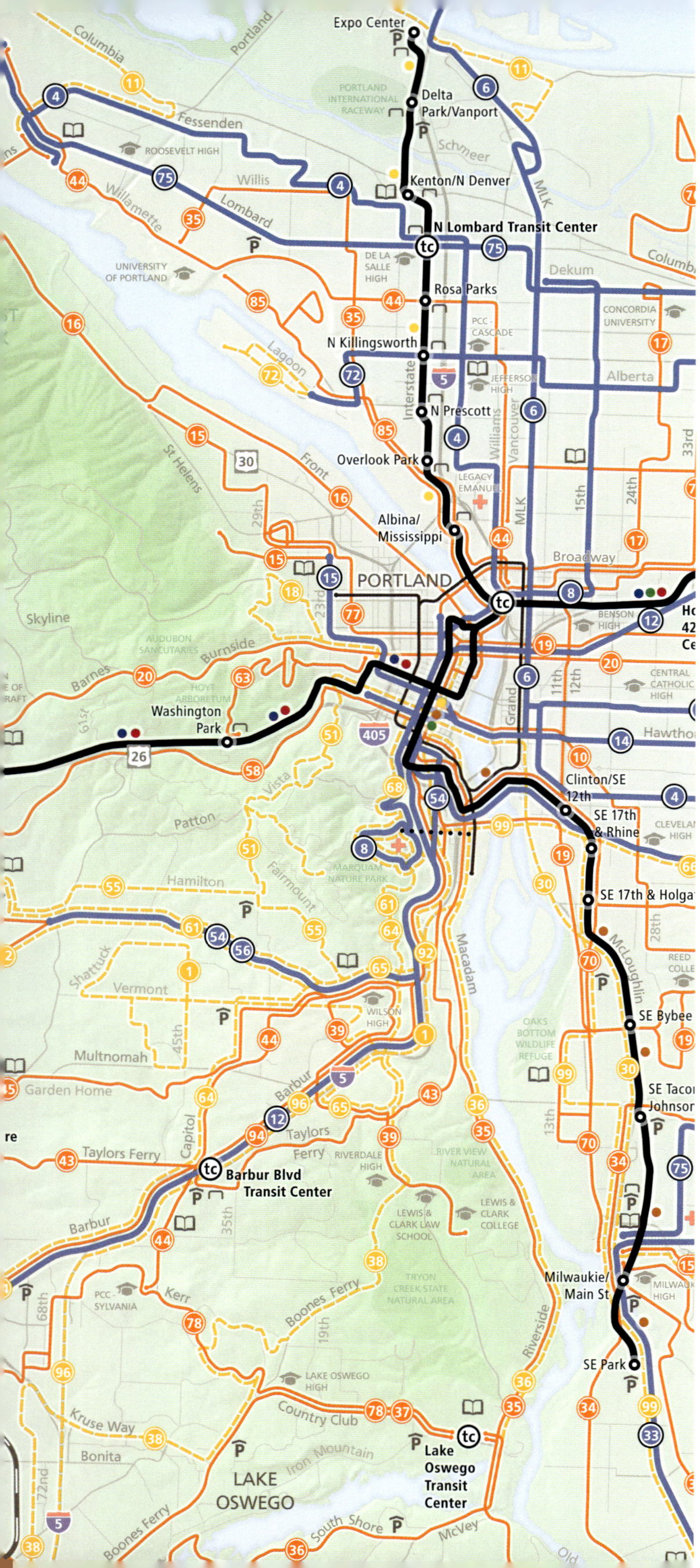

Unofficial TriMet System Map »

Grant Humphries

This bus and rail system map was created to be distributed to users of TriMet's public transit system in Portland, Oregon, to provide an understanding of the core properties of each transit line. In addition to displaying where each route goes, I sought to illustrate the mode, frequency, span, and connections among them.

Knowing when and how often transit service runs is important to transit riders because that information means the difference between a quick, easy trip and a long, difficult one[1]. To clearly display this information, I used contrasting hues and kept the saturation of all routes high to pop against the background. I used a spectrum of heavy, solid lines representing frequent service and lightweight, dashed lines portraying rush-hour-only service. Hue, value, line weight, and dashing also differentiate among various transit modes, including buses, light rail, and streetcars.

A central challenge of this project was accommodating the map's scale, which was required to be set at 1:100,000. I used a number of generalization techniques[2] to modify the data in order to effectively display it. To address routes' circuitous sections I used selection, removing nonessential segments in order to focus on the essence of the route. In cases where a transit line utilized both parts of a couplet, creating the appearance of a single feature with a weight heavier than the rest of the route, I combined the pair and labeled the street names above and below the feature. To clarify the space between sections that doubled back on themselves, I used displacement to eliminate any overlap and smoothed all of the routes to improve their aesthetic appeal. I used offsets, or displacement, to handle transit lines with common street segments that made them overlap. I wanted each route to have a distinct, unimpeded geometry whenever possible.

[1] *http://humantransit.org/2010/08/basics-the-case-for-frequency-mapping.html*

[2] *https://en.wikipedia.org/wiki/Cartographic_generalization#Methods*

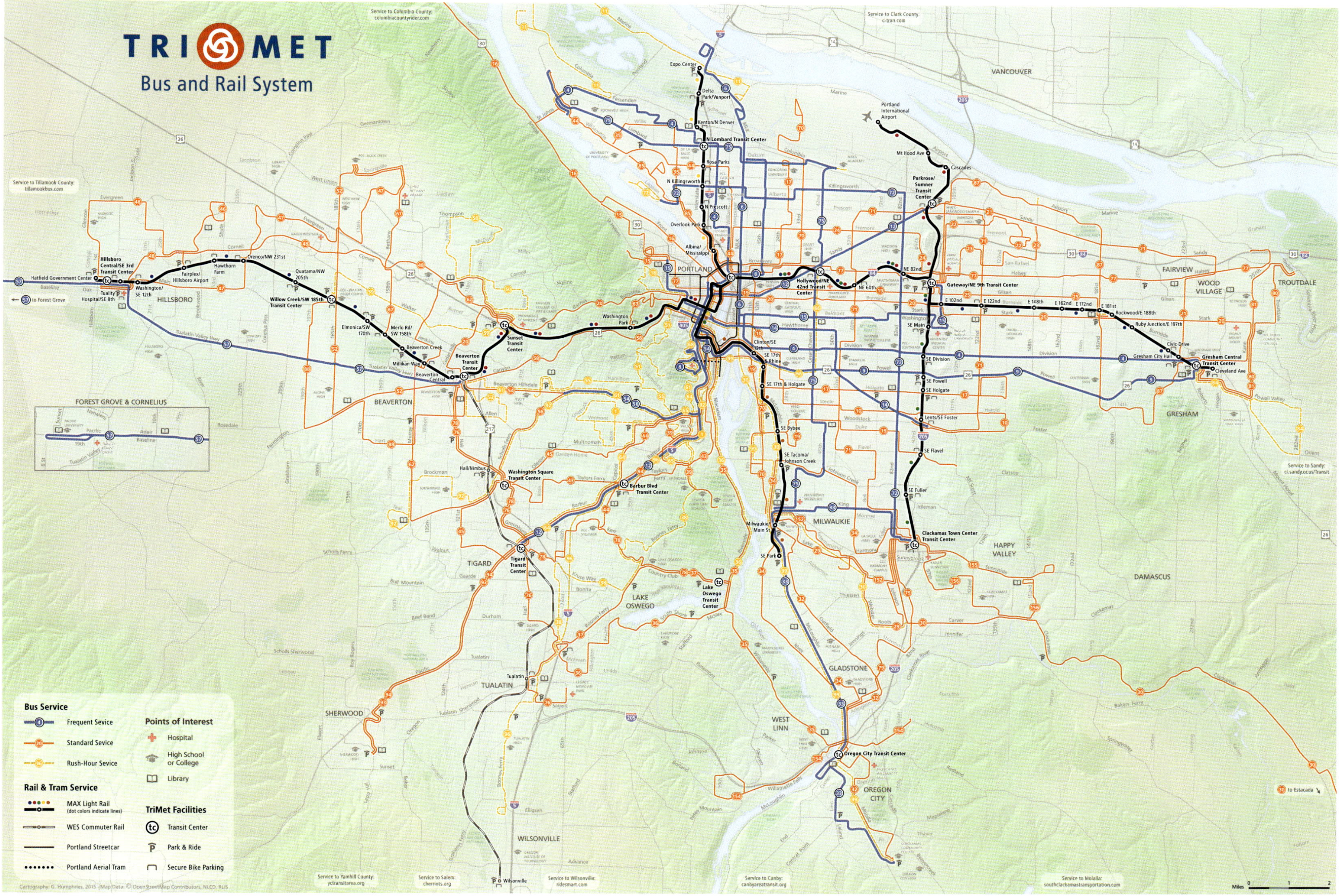
TRIMET
Bus and Rail System
Bus Service
Frequent Sevice
Standard Sevice
Rush-Hour Sevice
Points of Interest
Hospital
High School or College
Library
Rail & Tram Service
MAX Light Rail (dot colors indicate lines)
WES Commuter Rail
Portland Streetcar
Portland Aerial Tram
TriMet Facilities
Transit Center
Park & Ride
Secure Bike Parking
FOREST GROVE & CORNELIUS
Service to Columbia County: columbiacountyrider.com
Service to Clark County: c-tran.com
Service to Tillamook County: tillamookbus.com
Service to Yamhill County: yctransitarea.org
Service to Salem: cherriots.org
Service to Wilsonville: ridesmart.com
Service to Canby: canbyareatransit.org
Service to Molalla: southclackamastransportation.com
Service to Sandy: ci.sandy.or.us/Transit
to Forest Grove
to Estacada
VANCOUVER
PORTLAND
HILLSBORO
BEAVERTON
TIGARD
LAKE OSWEGO
MILWAUKIE
HAPPY VALLEY
DAMASCUS
GRESHAM
FAIRVIEW
WOOD VILLAGE
TROUTDALE
GLADSTONE
WEST LINN
OREGON CITY
TUALATIN
SHERWOOD
WILSONVILLE
FOREST PARK
Expo Center
Delta Park/Vanport
Kenton/N Denver
N Lombard Transit Center
Rosa Parks
N Killingsworth
N Prescott
Overlook Park
Albina/Mississippi
Portland International Airport
Mt Hood Ave
Cascades
Parkrose/Sumner Transit Center
Gateway/NE 9th Transit Center
Hollywood/NE 42nd Transit Center
NE 60th
NE 82nd
Hatfield Government Center
Hillsboro Central/SE 3rd Transit Center
Washington/SE 12th
Tuality Hospital/SE 8th
Fairplex/Hillsboro Airport
Hawthorn Farm
Orenco/NW 231st
Quatama/NW 205th
Willow Creek/SW 185th Transit Center
Elmonica/SW 170th
Merlo Rd/SW 158th
Beaverton Creek
Millikan Way
Beaverton Central
Beaverton Transit Center
Sunset Transit Center
Washington Park
Clinton/SE 12th
SE 17th & Rhine
SE 17th & Holgate
SE Bybee
SE Tacoma/Johnson Creek
Milwaukie/Main St
SE Park
E 102nd
E 122nd
E 148th
E 162nd
E 172nd
E 181st
Rockwood/E 188th
Ruby Junction/E 197th
Civic Drive
Gresham City Hall
Gresham Central Transit Center
Cleveland Ave
SE Main
SE Division
SE Powell
SE Holgate
Lents/SE Foster
SE Flavel
SE Fuller
Clackamas Town Center Transit Center
Hall/Nimbus
Washington Square Transit Center
Barbur Blvd Transit Center
Tigard Transit Center
Lake Oswego Transit Center
Oregon City Transit Center
Tualatin
Wilsonville
Miles 0 1 2
Cartography: G. Humphries, 2015 · Map Data: © OpenStreetMap Contributors, NLCD, RLIS

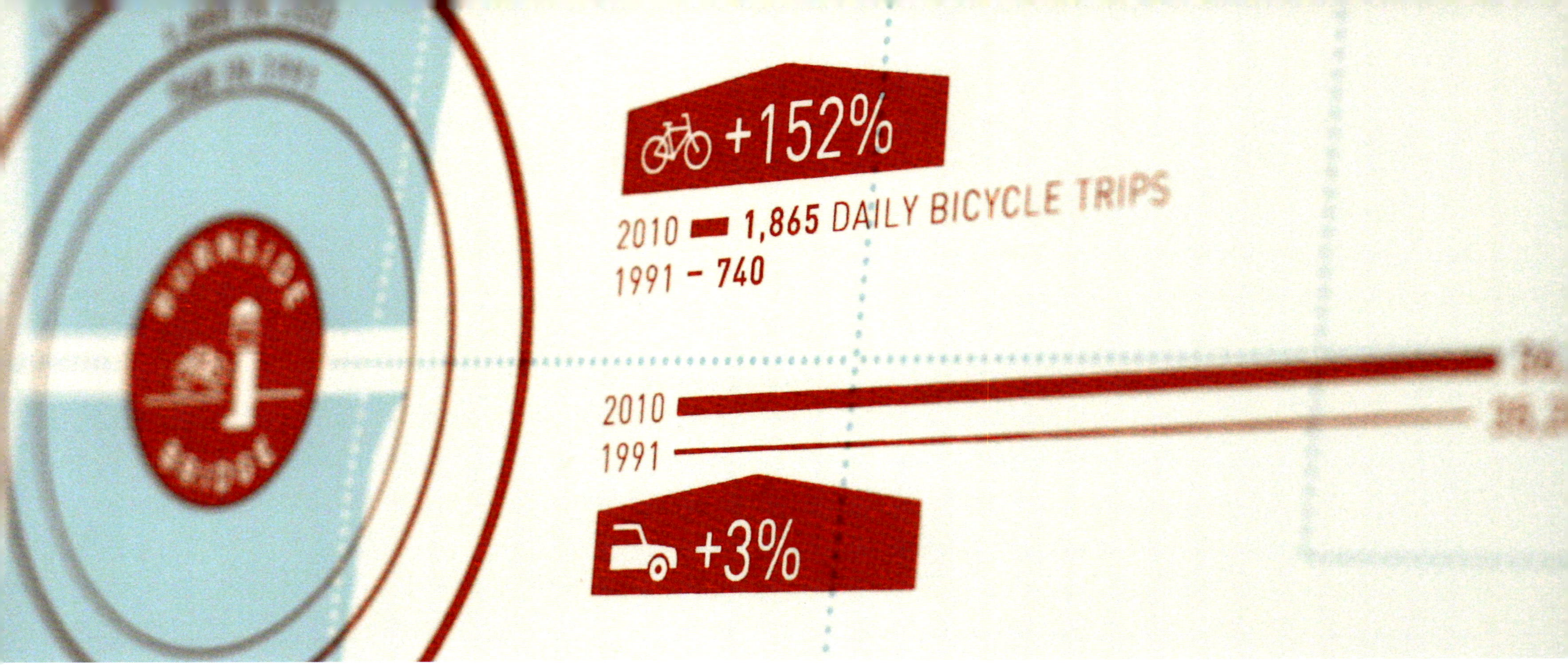

Bicycletown/Bridgetown »

Ryan Sullivan | Paste in Place

Portland's bridges are among the busiest bicycle routes in the city. The Bureau of Transportation has conducted bicycle counts for the Broadway Bridge, Steel Bridge, Burnside Bridge, and Hawthorne Bridge since 1991. This infographic map displays data from the bicycle counts in order to compare the ridership levels among bridges, show growth in ridership over the past two decades, and reveal changes in automobile and bicycle use of the bridges.

We designed this poster for an art show. Its bright colors and bold form encourage visitors to approach it. The data presented in the map offers visitors an opportunity to explore and share their own insights about cycling in Portland.

To supplement the data-driven aspects of the graphic, we designed a logomark to mark the location of each bridge. The logomarks integrate salient characteristics of each bridge with a bicycle symbol. We also produced the logomarks as buttons and distributed them along with the poster to encourage civic pride and dialogue, asking commuters, "Which bridge do you bike?"

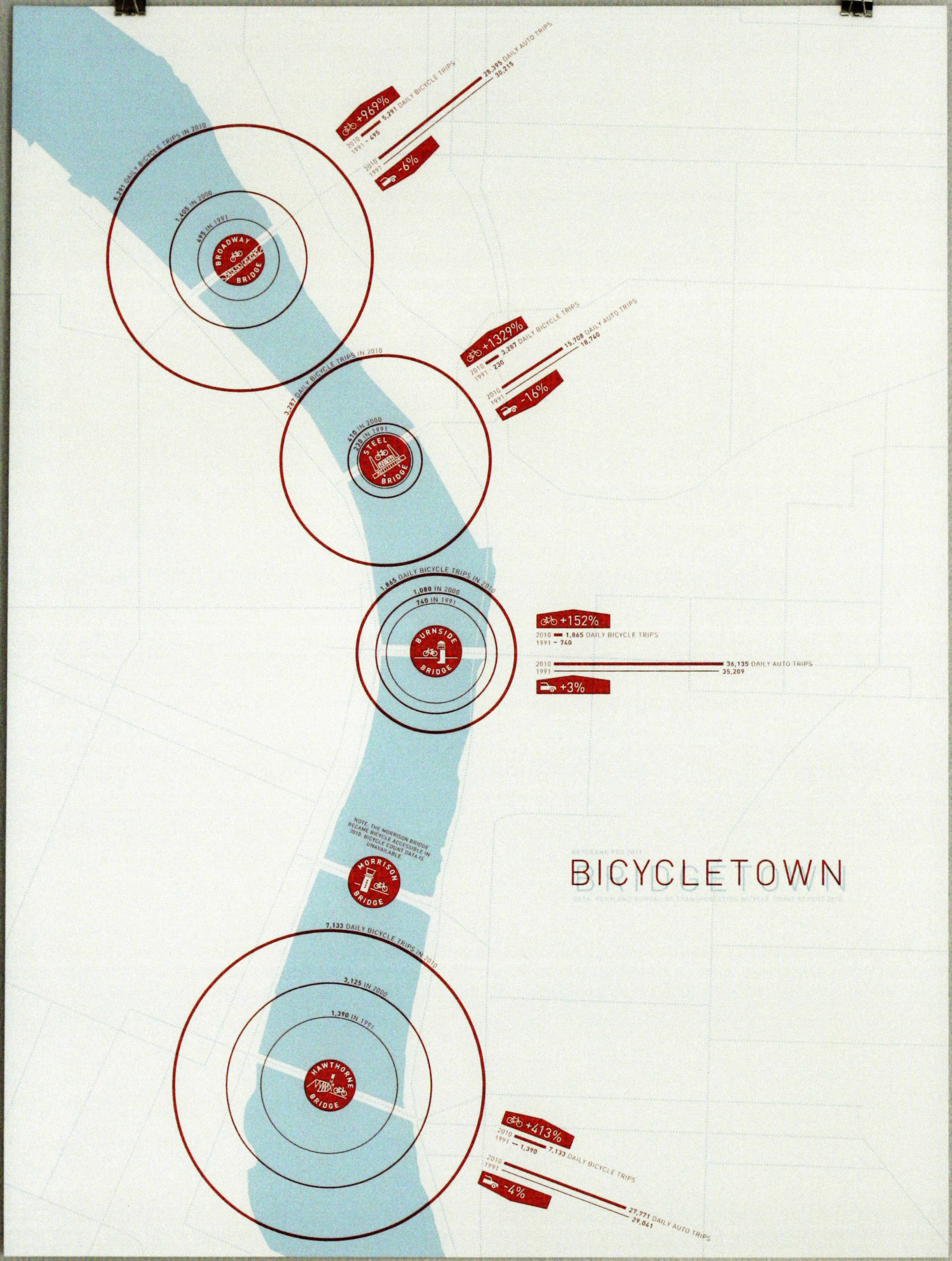
BICYCLETOWN
BROADWAY BRIDGE
495 IN 1991
1,405 IN 2000
5,291 DAILY BICYCLE TRIPS IN 2010
+969%
2010 — 5,291 DAILY BICYCLE TRIPS
1991 – 495
2010
1991
28,395 DAILY AUTO TRIPS
30,215
-6%
STEEL BRIDGE
230 IN 1991
410 IN 2000
3,287 DAILY BICYCLE TRIPS IN 2010
+1329%
2010 — 3,287 DAILY BICYCLE TRIPS
1991 – 230
2010
1991
15,708 DAILY AUTO TRIPS
18,740
-16%
BURNSIDE BRIDGE
740 IN 1991
1,080 IN 2000
1,865 DAILY BICYCLE TRIPS IN 2010
+152%
2010 — 1,865 DAILY BICYCLE TRIPS
1991 – 740
2010
1991
36,135 DAILY AUTO TRIPS
35,209
+3%
NOTE: THE MORRISON BRIDGE BECAME BICYCLE ACCESSIBLE IN 2010. BICYCLE COUNT DATA IS UNAVAILABLE.
MORRISON BRIDGE
HAWTHORNE BRIDGE
1,390 IN 1991
3,125 IN 2000
7,133 DAILY BICYCLE TRIPS IN 2010
+413%
2010 — 7,133 DAILY BICYCLE TRIPS
1991 — 1,390
2010
1991
27,771 DAILY AUTO TRIPS
29,041
-4%

BOB MARSHALL COUNTRY

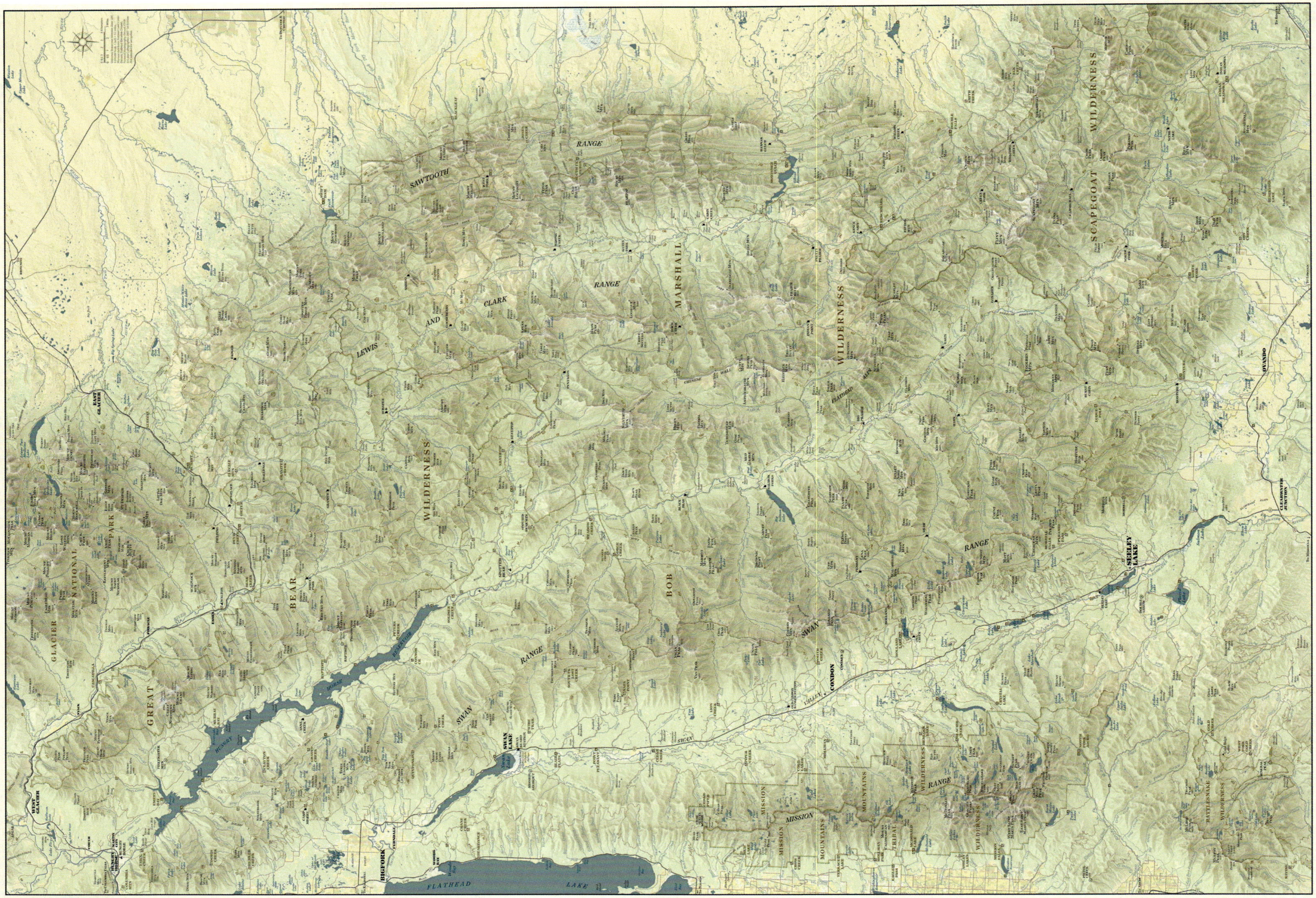

« Bob Marshall Country

Cairn Cartographics

The Bob Marshall Wilderness Complex is a vast, rugged area located in northwestern Montana. Affectionately called "The Bob," this wild landscape inspires a sense of awe and reverence in those who know it well.

The Bob is named for a mapmaker and conservationist who trekked across the Adirondacks, the northern Rockies, and as far as Alaska's Brooks Range. Bob Marshall was a tireless advocate for protecting and preserving wild places. Wherever he went, he drew maps so he later could call to mind the land he had walked.

This map aims to honor Marshall by transporting people who have visited The Bob back to those mountain peaks, rushing rivers, and limestone walls. We imagine friends gathered around the map on a cold winter night telling stories of past trips and scheming future adventures.

As a condensed version of our popular recreation maps, we wanted this map to be detailed enough to find a trail, a small creek, or a specific cabin—things commonly used as landmarks in the backcountry—but focused primarily on accentuating the dramatic natural features and terrain. Many cultural features are visible, but they are styled in a way that does not distract from the natural topography. The only political boundaries included are federally designated Wilderness areas, including three that make up the Bob Marshall Wilderness Complex.

Terrain texturing highlights iconic reefs and cliffs that tower above pristine waterways. Natural tones are used to style the land cover data so that it's depicted accurately and blends well with the rest of the map. A serifed typeface complements the natural landscape throughout. Our true goal for this map is to draw viewers deeply into the detailed, beautiful terrain so they will recall specific landmarks from the wild country they've been lucky enough to travel.

We imagine friends gathered around the map on a cold winter night, telling stories of past trips and scheming future adventures.

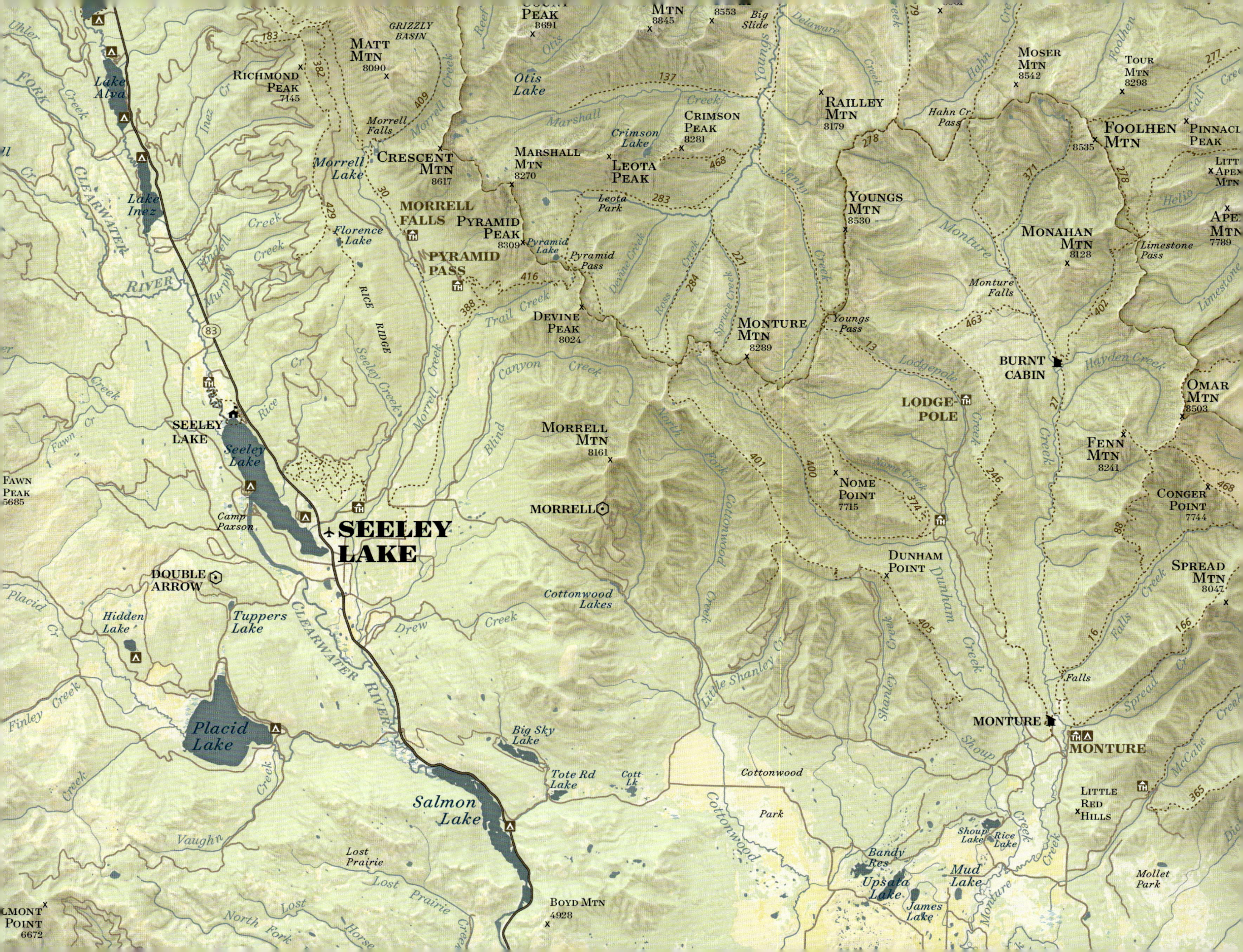

SEELEY LAKE
Seeley Lake
SEELEY LAKE
Lake Alva
Lake Inez
CLEARWATER RIVER
FORK
Uhler
Inez Cr
Richmond Peak 7145
Matt Mtn 8090
Grizzly Basin
Morrell Falls
Morrell Lake
Crescent Mtn 8617
Morrell Falls
Marshall Mtn 8270
Crimson Lake
Crimson Peak 8281
Leota Peak
Leota Park
Otis Lake
Marshall Creek
Pyramid Peak 8309
Pyramid Lake
Pyramid Pass
Pyramid Pass
Florence Lake
Rice Ridge
Devine Peak 8024
Trail Creek
Canyon Creek
Seeley Creek
Morrell Creek
Morrell Mtn 8161
Morrell
Blind
Monture Mtn 8289
Youngs Pass
Youngs Mtn 8530
Railley Mtn 8179
Hahn Cr Pass
Moser Mtn 8542
Tour Mtn 8298
Foolhen Mtn 8535
Pinnacle Peak
Monahan Mtn 8128
Limestone Pass
Monture Falls
Burnt Cabin
Hayden Creek
Omar Mtn 8503
Lodgepole
Lodgepole
Fenn Mtn 8241
Nome Point 7715
Nome Creek
Conger Point 7744
Dunham Point
Dunham Creek
Spread Mtn 8047
Cottonwood Lakes
Cottonwood Creek
North Fork
Little Shanley Cr
Shanley Creek
Falls
Monture
Monture
Shoup
Little Red Hills
Fawn Peak 5685
Fawn Cr
Camp Paxson
Double Arrow
Placid Cr
Hidden Lake
Tuppers Lake
Clearwater River
Drew Creek
Placid Lake
Finley Creek
Big Sky Lake
Tote Rd Lake
Cott Lk
Cottonwood
Park
Salmon Lake
Vaughn
Lost Prairie
Lost Prairie
North Fork
Lost Horse
Boyd Mtn 4928
Bandy Res
Upsala Lake
Shoup Lake
Rice Lake
Mud Lake
James Lake
Mollet Park
Monture Creek
McCabe Creek
Spread Cr
Falls Creek
Delaware
Youngs Creek
Big Slide
Jenny
Murphy
Rice
83

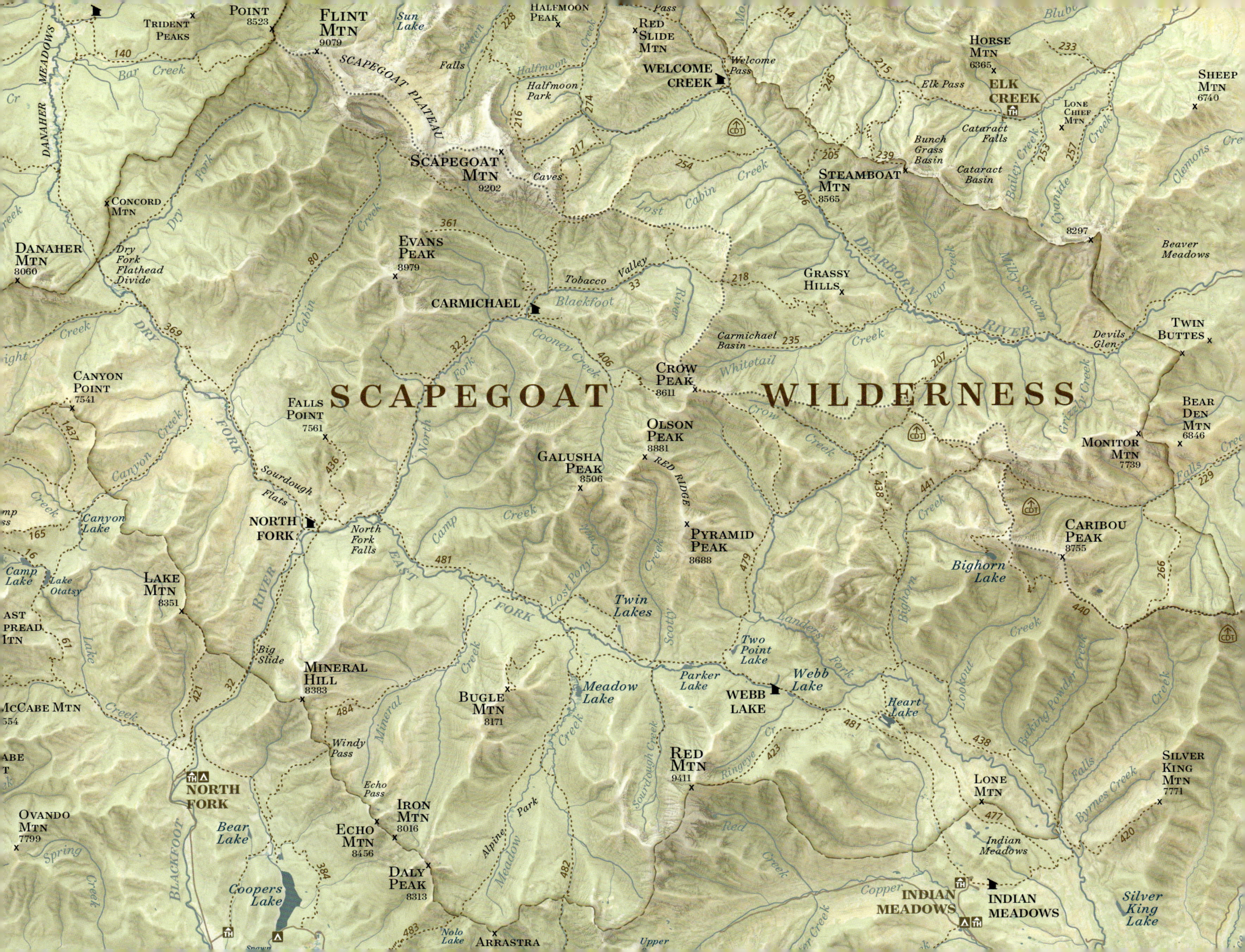

SCAPEGOAT WILDERNESS
Point 8523
Trident Peaks
Flint Mtn 9079
Sun Lake
Halfmoon Peak
Red Slide Mtn
Welcome Creek
Welcome Pass
Horse Mtn 6365
Elk Pass
Elk Creek
Sheep Mtn 6740
Lone Chief Mtn
Bar Creek
Meadows
Danaher
Scapegoat Plateau
Falls
Halfmoon
Halfmoon Park
Bunch Grass Basin
Cataract Falls
Cataract Basin
Scapegoat Mtn 9202
Caves
Steamboat Mtn 8565
Concord Mtn
Lost Cabin Creek
Dearborn River
Danaher Mtn 8060
Dry Fork Flathead Divide
Evans Peak 8979
Tobacco Valley
Carmichael
Blackfoot River
Grassy Hills
Beaver Meadows
8297
Pear Creek
Milky Stream
Devils Glen
Twin Buttes
Carmichael Basin
Cooney Creek
Whitetail Creek
Crow Peak 8611
Canyon Point 7541
Falls Point 7561
Dry Fork
Olson Peak 8881
Red Ridge
Crow Creek
Bear Den Mtn 6846
Monitor Mtn 7739
Galusha Peak 8506
Sourdough Flats
North Fork
North Fork Falls
Canyon Lake
Camp Creek
Pyramid Peak 8688
Caribou Peak 8755
Bighorn Lake
Camp Lake
Lake Otatsy
Lake Mtn 8351
East Fork
Lost Pony Creek
Twin Lakes
Scotty Creek
Landers Fork
Big Slide
Two Point Lake
Mineral Hill 8383
McCabe Mtn
Bugle Mtn 8171
Meadow Lake
Parker Lake
Webb Lake
Heart Lake
Lookout
Baking Powder Creek
Windy Pass
Echo Pass
North Fork
Red Mtn 9411
Ringeye Cr
Sourdough Creek
Lone Mtn
Silver King Mtn 7771
Byrnes Creek
Ovando Mtn 7799
Bear Lake
Iron Mtn 8016
Echo Mtn 8456
Alpine Park
Meadow Creek
Indian Meadows
Daly Peak 8313
Coopers Lake
Blackfoot
Spring Creek
Nolo Lake
Arrastra
Copper Creek
Indian Meadows
Indian Meadows
Silver King Lake

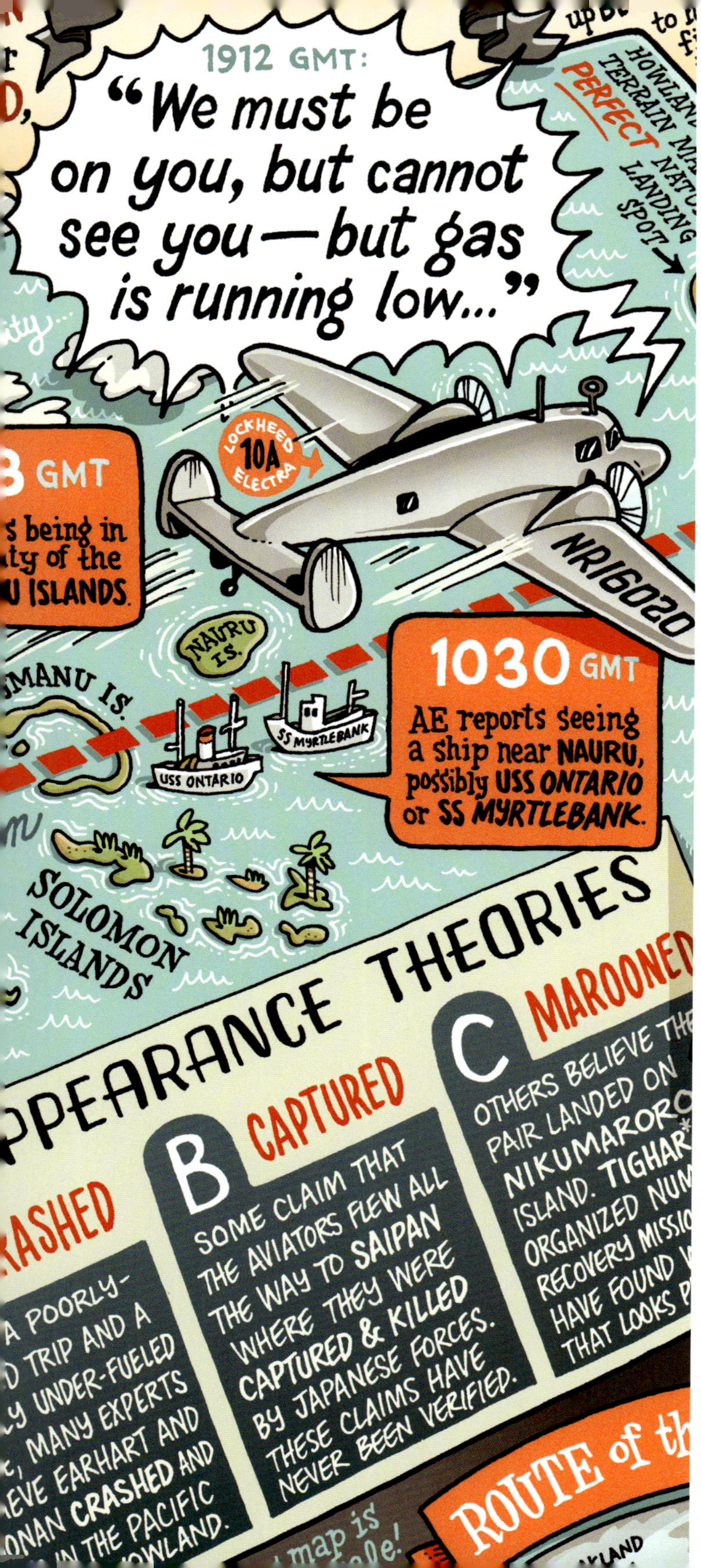

Amelia Earhart's Final Flight »

Kevin Cannon

Some of my favorite childhood memories occurred when peering down on the world from an airplane, marveling at not only how small everything looked, but also how connected. When seen from the air, buildings, rivers, and soccer fields were linked in a way I had never experienced before. I credit these moments in childhood—when I began looking at and understanding the world in a new way—with changing my life. I now look at history the same way: from above. I believe historical events make the most sense when viewed from a distance, allowing the historian (or in my case, armchair historian) to see a complex, interconnected network of people, places, and events.

The allure of this map of Amelia Earhart's Final Flight—and perhaps all maps—is that it allows the viewer to enter the image at any point and explore. Free from the rigidity of a written narrative, the reader can jump in and out, circling around and discovering new bits of information with each pass.

I knew little about this historic flight before I began to produce the map. I knew that Earhart's plane had disappeared somewhere far away while she was trying to find a small island and that search-and-rescue attempts had come up short. Drawing this map was an excuse to learn more and make sense of Amelia Earhart's and Fred Noonan's famous last flight. I couldn't fit every bit of information I learned into the map because of limits on the size of the page, but I hope this drawing sets the stage, providing an image of the interwoven strands of this complicated story, whetting the reader's appetite for further exploration.

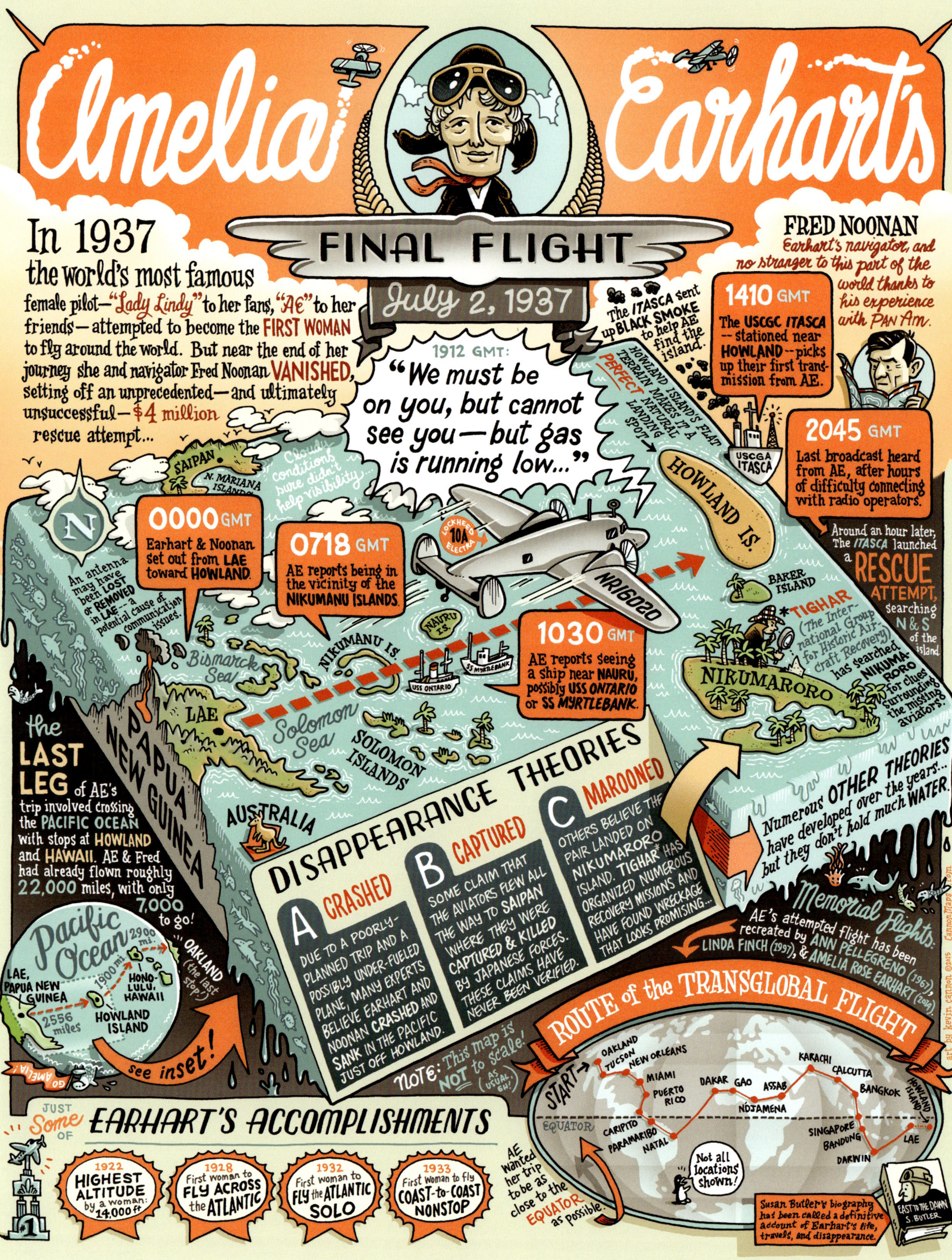

Amelia Earhart's
FINAL FLIGHT
July 2, 1937
In 1937 the world's most famous female pilot—"Lady Lindy" to her fans, "AE" to her friends—attempted to become the FIRST WOMAN to fly around the world. But near the end of her journey she and navigator Fred Noonan VANISHED, setting off an unprecedented—and ultimately unsuccessful—$4 million rescue attempt...
FRED NOONAN
Earhart's navigator, and no stranger to this part of the world thanks to his experience with PAN AM.
1912 GMT:
"We must be on you, but cannot see you—but gas is running low..."
The ITASCA sent up BLACK SMOKE to help AE find the island.
1410 GMT
The USCGC ITASCA -- stationed near HOWLAND -- picks up their first transmission from AE.
2045 GMT
Last broadcast heard from AE, after hours of difficulty connecting with radio operators.
Around an hour later, The ITASCA launched a RESCUE ATTEMPT, searching N&S of the island.
HOWLAND ISLAND'S FLAT TERRAIN MAKES IT A PERFECT NATURAL LANDING SPOT
HOWLAND IS.
USCGA ITASCA
BAKER ISLAND
SAIPAN
N. MARIANA ISLANDS
Cloudy conditions sure didn't help visibility
0000 GMT
Earhart & Noonan set out from LAE toward HOWLAND.
0718 GMT
AE reports being in the vicinity of the NIKUMANU ISLANDS.
1030 GMT
AE reports seeing a ship near NAURU, possibly USS ONTARIO or SS MYRTLEBANK.
An antenna may have been LOST or REMOVED in LAE—a potential cause of communication issues.
LOCKHEED 10A ELECTRA
NR16020
NAURU IS.
USS ONTARIO
SS MYRTLEBANK
NIKUMANU IS.
Bismarck Sea
LAE
Solomon Sea
SOLOMON ISLANDS
PAPUA NEW GUINEA
AUSTRALIA
* TIGHAR (The International Group for Historic Aircraft Recovery) has searched NIKUMARORO for clues surrounding the missing aviators.
NIKUMARORO
the LAST LEG of AE's trip involved crossing the PACIFIC OCEAN with stops at HOWLAND and HAWAII. AE & Fred had already flown roughly 22,000 miles, with only 7,000 to go!
DISAPPEARANCE THEORIES
A CRASHED
DUE TO A POORLY-PLANNED TRIP AND A POSSIBLY UNDER-FUELED PLANE, MANY EXPERTS BELIEVE EARHART AND NOONAN CRASHED AND SANK IN THE PACIFIC JUST OFF HOWLAND.
B CAPTURED
SOME CLAIM THAT THE AVIATORS FLEW ALL THE WAY TO SAIPAN WHERE THEY WERE CAPTURED & KILLED BY JAPANESE FORCES. THESE CLAIMS HAVE NEVER BEEN VERIFIED.
C MAROONED
OTHERS BELIEVE THE PAIR LANDED ON NIKUMARORO ISLAND. TIGHAR* HAS ORGANIZED NUMEROUS RECOVERY MISSIONS AND HAVE FOUND WRECKAGE THAT LOOKS PROMISING...
Numerous OTHER THEORIES have developed over the years... but they don't hold much WATER.
Memorial Flights:
AE's attempted flight has been recreated by ANN PELLEGRENO (1967), LINDA FINCH (1997), & AMELIA ROSE EARHART (2014).
Pacific Ocean
LAE, PAPUA NEW GUINEA
2556 miles
HOWLAND ISLAND
1900 mi.
HONOLULU, HAWAII
2900 mi.
OAKLAND (the last stop)
see inset!
GO AMELIA!
NOTE: This map is NOT to scale! (AS USUAL, EH!)
ROUTE of the TRANSGLOBAL FLIGHT
START
OAKLAND
TUCSON
NEW ORLEANS
MIAMI
PUERTO RICO
CARIPITO
PARAMARIBO
NATAL
DAKAR
GAO
NDJAMENA
ASSAB
KARACHI
CALCUTTA
BANGKOK
SINGAPORE
BANDUNG
DARWIN
LAE
HOWLAND ISLAND
EQUATOR
Not all locations shown!
AE wanted her trip to be as close to the EQUATOR as possible.
Susan Butler's biography has been called a definitive account of Earhart's life, travels, and disappearance.
EAST TO THE DAWN S. BUTLER
JUST Some OF EARHART'S ACCOMPLISHMENTS
1922 HIGHEST ALTITUDE by a woman: 14,000 ft.
1928 First woman to FLY ACROSS the ATLANTIC
1932 First woman to FLY the ATLANTIC SOLO
1933 First woman to fly COAST-to-COAST NONSTOP
art by kevin cannon ©2015 CannonMaps.com

A.A. Mäkijärvi proudly presents:

The Magnificent BEARS of the Glorious Nation of FINLAND

Approximately before & after the year 2010

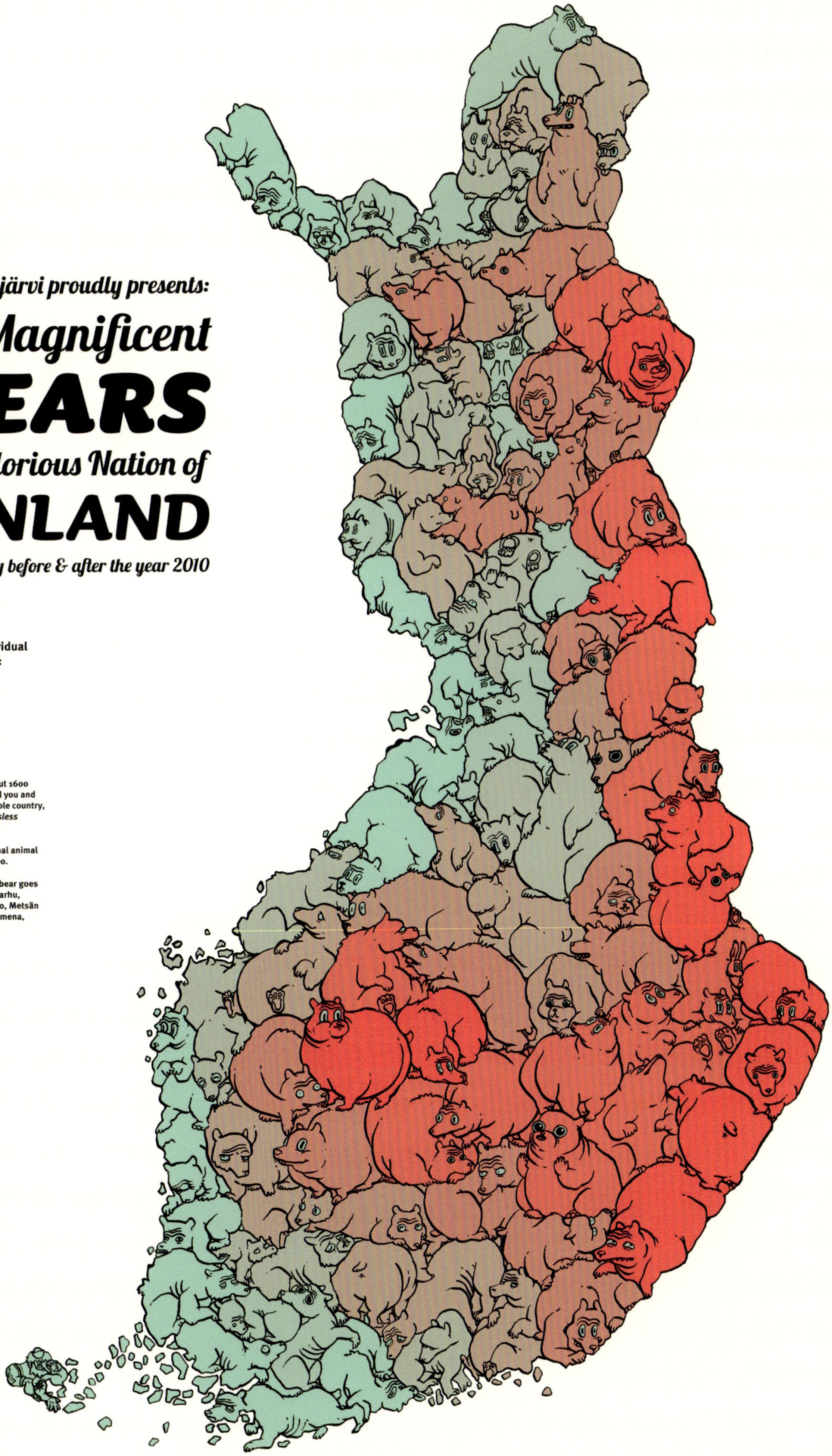

The amount of individual bears per 1000 km²:

- 0 - 2,0
- 2,1 - 4,0
- 4,1 - 6,0
- 6,1 - ∞

Which means there is about 1600 graceful bears ready to kill you and your loved ones in the whole country, excluding the *Ursus arctosless* province of Ahvenanmaa.

Fun fact! Bear is the national animal of Finland. And Russia's too.

In the Finnish language a bear goes by the following names: Karhu, Mesikämmen, Otso, Kontio, Metsän kuningas, Nalle, Metsän omena, Ohto and Nallukka.

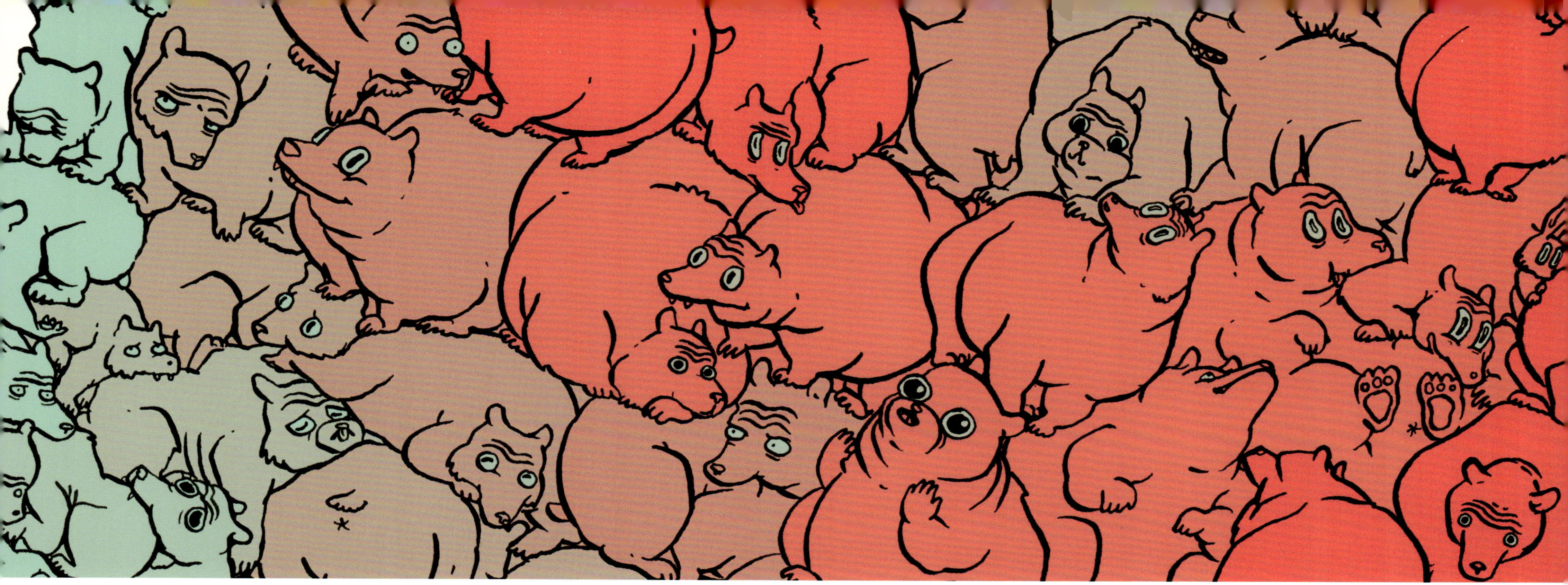

« Bears of Finland

Annukka Mäkijärvi

As a challenge, in creating an infographic map of Finland, I decided to illustrate one particular subject by making the subject itself the variable. This approach creates an intimate link between the form and the content. I chose to make a Finland-shaped mass of bears that would show the number of bears in the country represented by illustrations of bears, in an effective and easy-to-understand format. My design goal was to combine illustration and infographic in a way that would support each equally.

In the planning process a humorous and unconventional approach felt right. I was eager to find out how far I could take a nontraditional, somewhat silly map illustration without sacrificing information and clarity. A viewer should be able to explore and understand the map effortlessly, regardless of their age or background.

I chose bright, aesthetically pleasing shades for the color scheme, with red marking greater numbers of bears, and green fewer. I made sure that a viewer could understand the map just through the coloring, without the outlines of the bears. On the other hand, one can also read the map without colors, by interpreting the body shapes of the bears. But I think more is more, so the combination of outline drawings and colors produces the most expressive illustration.

The drawing of the bears prompted important questions: How could I show readers bear population density? What about regions that have no bears at all? I decided to draw the bears in bear-dense areas as excessively chubby and joyful, while bears in less densely bear-populated zones are emaciated, deformed, miserable creatures. Finland's rough and jagged west coast presented an especially challenging shape for the bears' outlines.

Finland's only bear-free area is Ahvenanmaa, the southwestern islands. There I drew a single sailor, as a visual pun. In Finnish, sailors are nicknamed *merikarhu*, which translates to "sea bear."

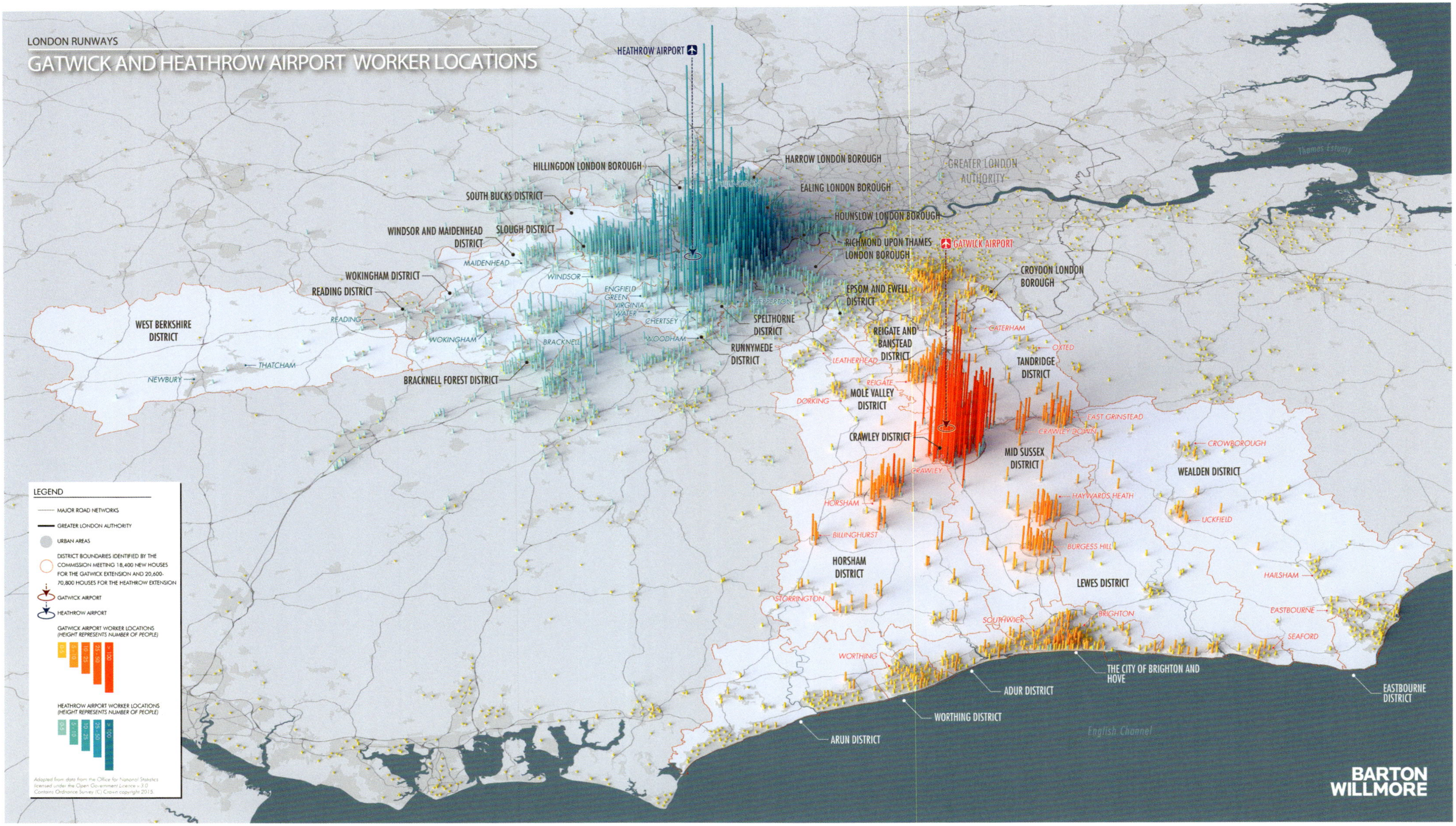

LONDON RUNWAYS
GATWICK AND HEATHROW AIRPORT WORKER LOCATIONS
HEATHROW AIRPORT
GATWICK AIRPORT
HILLINGDON LONDON BOROUGH
HARROW LONDON BOROUGH
EALING LONDON BOROUGH
SOUTH BUCKS DISTRICT
HOUNSLOW LONDON BOROUGH
WINDSOR AND MAIDENHEAD DISTRICT
SLOUGH DISTRICT
RICHMOND UPON THAMES LONDON BOROUGH
GREATER LONDON AUTHORITY
Thames Estuary
CROYDON LONDON BOROUGH
WOKINGHAM DISTRICT
READING DISTRICT
EPSOM AND EWELL DISTRICT
MAIDENHEAD
WINDSOR
ENGFIELD GREEN
VIRGINIA WATER
CHERTSEY
WOODHAM
SPELTHORNE DISTRICT
RUNNYMEDE DISTRICT
WEST BERKSHIRE DISTRICT
READING
WOKINGHAM
BRACKNELL
THATCHAM
NEWBURY
BRACKNELL FOREST DISTRICT
REIGATE AND BANSTEAD DISTRICT
CATERHAM
OXTED
TANDRIDGE DISTRICT
LEATHERHEAD
REIGATE
DORKING
MOLE VALLEY DISTRICT
EAST GRINSTEAD
CRAWLEY DOWN
CRAWLEY DISTRICT
CRAWLEY
MID SUSSEX DISTRICT
CROWBOROUGH
WEALDEN DISTRICT
HORSHAM
HAYWARDS HEATH
UCKFIELD
BILLINGHURST
BURGESS HILL
HORSHAM DISTRICT
LEWES DISTRICT
HAILSHAM
STORRINGTON
SOUTHWICK
BRIGHTON
EASTBOURNE
SEAFORD
WORTHING
THE CITY OF BRIGHTON AND HOVE
ADUR DISTRICT
EASTBOURNE DISTRICT
WORTHING DISTRICT
ARUN DISTRICT
English Channel
LEGEND
MAJOR ROAD NETWORKS
GREATER LONDON AUTHORITY
URBAN AREAS
DISTRICT BOUNDARIES IDENTIFIED BY THE COMMISSION MEETING 18,400 NEW HOUSES FOR THE GATWICK EXTENSION AND 20,600-70,800 HOUSES FOR THE HEATHROW EXTENSION
GATWICK AIRPORT
HEATHROW AIRPORT
GATWICK AIRPORT WORKER LOCATIONS (HEIGHT REPRESENTS NUMBER OF PEOPLE)
0-5
5-10
10-25
25-50
> 100
HEATHROW AIRPORT WORKER LOCATIONS (HEIGHT REPRESENTS NUMBER OF PEOPLE)
0-5
5-10
10-25
25-50
> 100
Adapted from data from the Office for National Statistics licensed under the Open Government Licence v.3.0 Contains Ordnance Survey (C) Crown copyright 2015.
BARTON WILLMORE

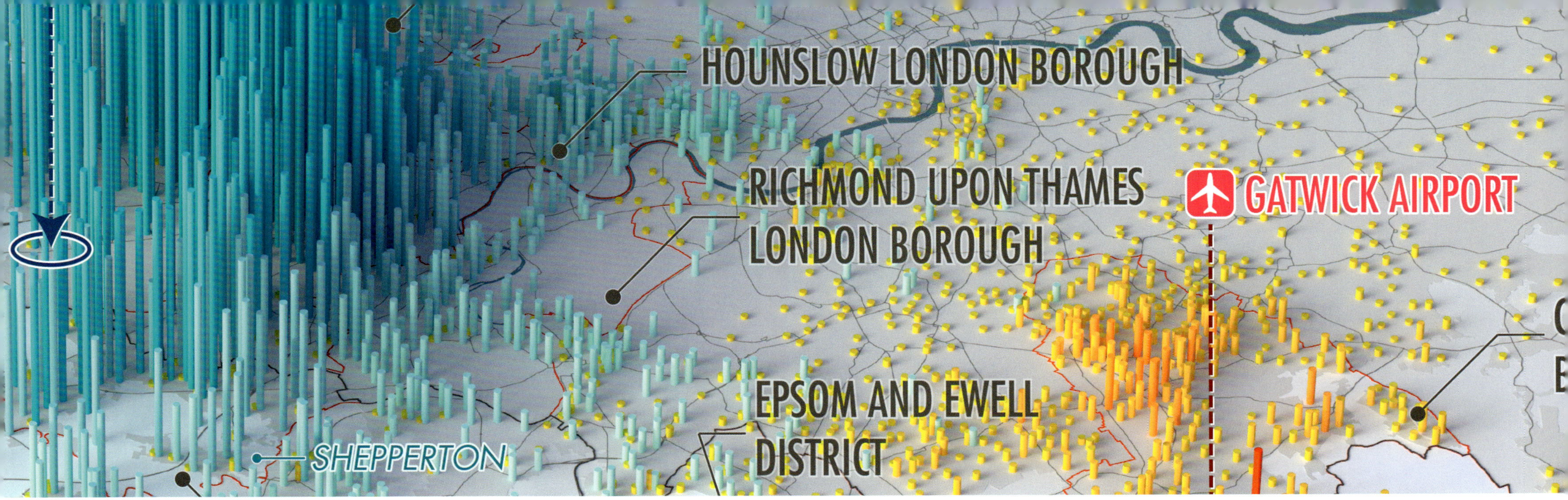

« Gatwick and Heathrow Worker Locations

Craig Taylor

The alchemy of turning raw data into meaningful maps is what excites me most. I've started taking a few risks as I strive to create interesting cartography and infographics. For this map I learned how to use new software and new techniques in lighting, modeling, and color to render a visualization that would catch people's attention. I think the risk paid off.

The expansion of UK airports always raises some consternation among the NIMBY community and environmentalists. But the economic, social, and employment prospects for local areas are among the most important questions to consider in the planning process. After the Sir Howard Davies Airports Commission concluded that either Heathrow or Gatwick Airport would require expansion, I decided to create a visualization to aid in assessing the current employment situation and the impact of each airport's enlargement.

This visualization analyzes the current spatial variance in travel-to-work patterns between residence and workplace for the two airports to help define the area of greatest impact based on employment, distance to work, and housing needs. Visualizing the flow data in 3-D generates a real and immediate sense of distribution across the workforce. Heathrow shows a smooth, typical spatial variation and flow rate in the workforce, but the Gatwick flow rate peaks around the airport and then shows a more dispersed, sporadic, and far-reaching workforce.

I wanted the whole map to be bright, the fonts to be light and thin, and the basemap to be subtle enough to show the context of the urban form. Modeling in 3-D allowed me to use advanced HDR lighting to achieve a bright aesthetic that also allows subtle shadowing from the peaks, adding a level of realism to the overall visualization.

Using a dataset with such granular data provided a level of detail and density that allowed the extruded points to give the map texture, while color and height distinguish each workforce. All these elements used together ensure that the map visualizes the intensity of commuter flows effectively. Darker hues and taller peaks indicate a greater flow of workers from their residences and lighter, smaller peaks show lower flow volumes.

Network and commuter data is fascinating and ideal for visualizing in three dimensions. This modeling technique allows for a clearer, more concise insight than is possible with more traditional approaches.

Losing Ground

PRO PUBLICA · THE LENS

DONATE TO PROPUBLICA · DONATE TO THE LENS

PAST / PRESENT: 1922 | 2014 · AN ENGINEERED COAST: LEVEES | CANALS | OIL/GAS · AN UNCERTAIN FUTURE: SEVERE | MODERATE

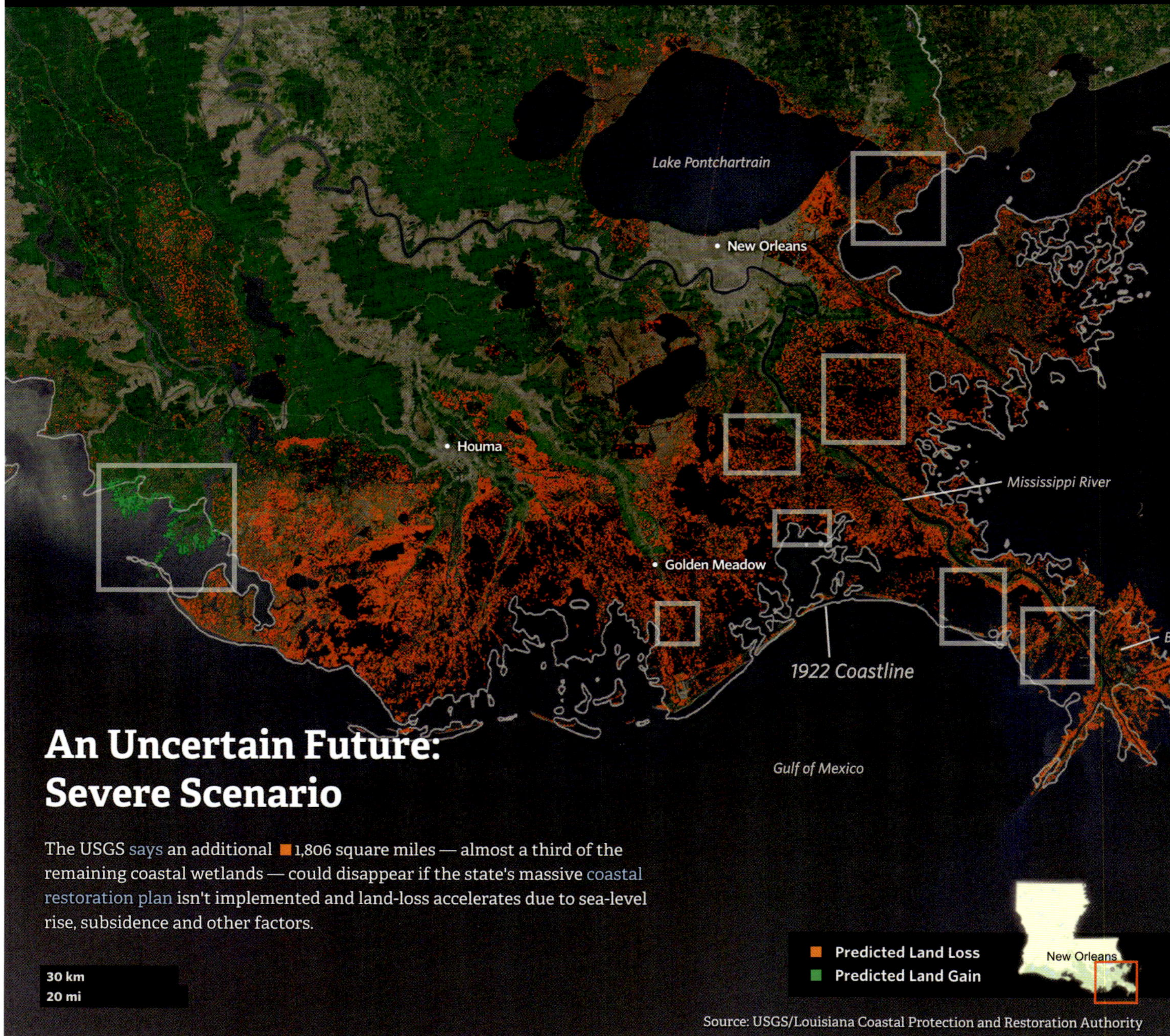

An Uncertain Future: Severe Scenario

The USGS says an additional 1,806 square miles — almost a third of the remaining coastal wetlands — could disappear if the state's massive coastal restoration plan isn't implemented and land-loss accelerates due to sea-level rise, subsidence and other factors.

Source: USGS/Louisiana Coastal Protection and Restoration Authority

Louisiana is drowning, quickly.

In just 80 years, some 2,000 square miles of its coastal landscape have turned to open water, wiping places off maps, bringing the Gulf of Mexico to the back door of New Orleans and posing a lethal threat to an energy and shipping corridor vital to the nation's economy.

And it's going to get worse, even quicker.

Scientists now say one of the greatest environmental and economic disasters in the nation's history is rushing toward a catastrophic conclusion over the next 50 years, so far unabated and largely unnoticed.

Explore Delacroix, La.

At the current rates that the sea is rising and land is sinking, National Oceanic and Atmospheric Administration scientists say by 2100 the Gulf of Mexico could rise as much as 4.3 feet across this landscape, which has an average elevation of about 3 feet. If that happens, everything outside the protective levees — most of Southeast Louisiana — would be underwater.

The effects would be felt far beyond bayou country. The region best known for its self-proclaimed motto *"laissez les bons temps rouler"* — let the good times roll — is one of the nation's economic linchpins.

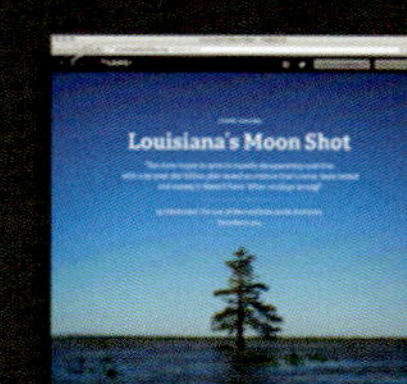

Louisiana's Moon Shot

This story is first in a

« Losing Ground

Al Shaw, Brian Jacobs (ProPublica), and Bob Marshall (The Lens)

The result of a year-long collaboration between two nonprofit newsrooms, "Losing Ground" is an interactive story about the slow-motion catastrophe taking place in southeast Louisiana. The region is losing about a football-field-size parcel of land every hour due to climate change, oil and gas exploration, and levees.

Combining satellite maps, historical maps, data, and on-the-ground, in-depth reporting and photography, "Losing Ground" allows readers to see where the land loss is happening, what is causing it, who is most affected, and what researchers predict the coastline will look like in the future.

To give readers a stark look at how precarious the coastline is today, we processed data from USGS/NASA's Landsat 8 satellite for our base map and introductory page. The result is an image that maintains the realistic greens of wetlands and whites of urban areas while accentuating open water intrusion into wetlands through exaggerated blues in places where sediment-rich water normally would be hard to distinguish from land.

From the introductory screen, readers step through a visual story that shows them what the Louisiana coastline used to look like and how human engineering led to coastal land loss. The story ends with a disturbing look at what the future may hold if we do nothing. By fading out the base map, adding an outline showing the extent of the coast in 1922 in white, and highlighting features like levees and canals in reds and oranges, we show that, more than any other factor, human development is responsible for the shape of the coastline today.

Readers can zoom in on eight different areas around the coast to read stories about how individual communities are dealing with land loss, and to see how the coastline has changed over time. Each one of these eight areas features an interactive timeline that adds or subtracts land from the map through the decades. Lacking satellite coverage in the 1930s, we relied instead on a USGS study of land change over the 20th century to build these views. The labels on the maps are tied to places mentioned in the story so readers can move back and forth on the timeline, and in many cases see features that existed earlier in the 20th century that are now submerged in large lakes or open water.

Photos and a richly reported story humanize the map with on-the-ground perspective. After reading the text and looking through the photographs, readers will understand the full import of a changing coastline on all these places—just in time to watch them fade away.

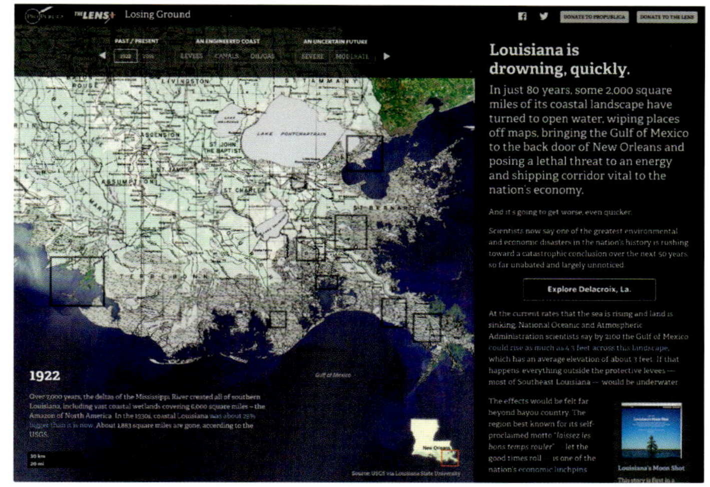

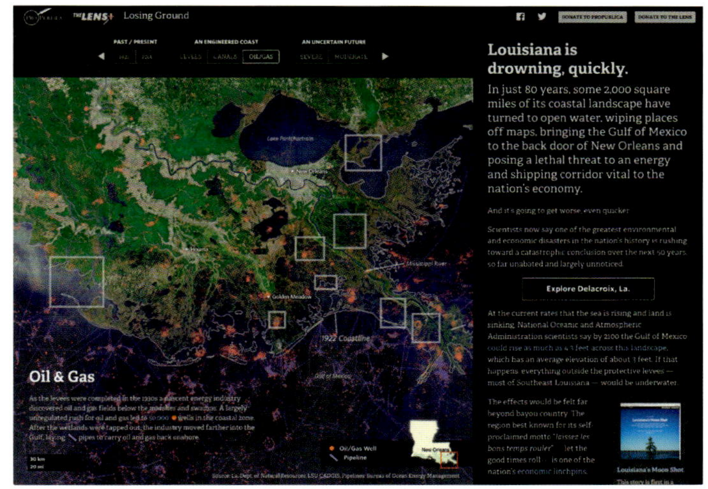

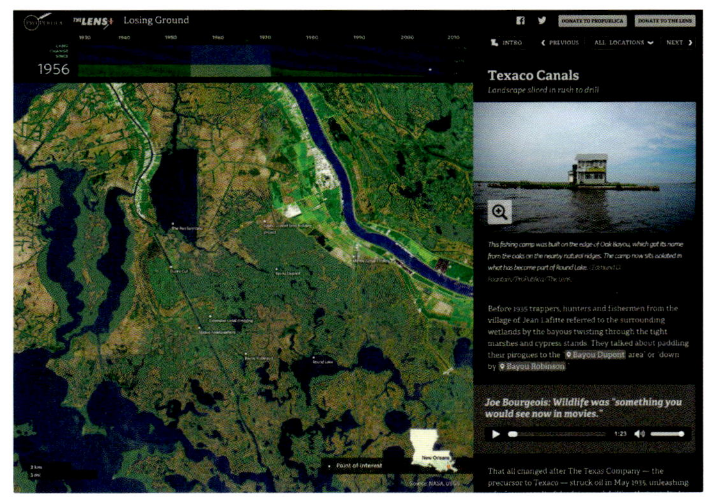

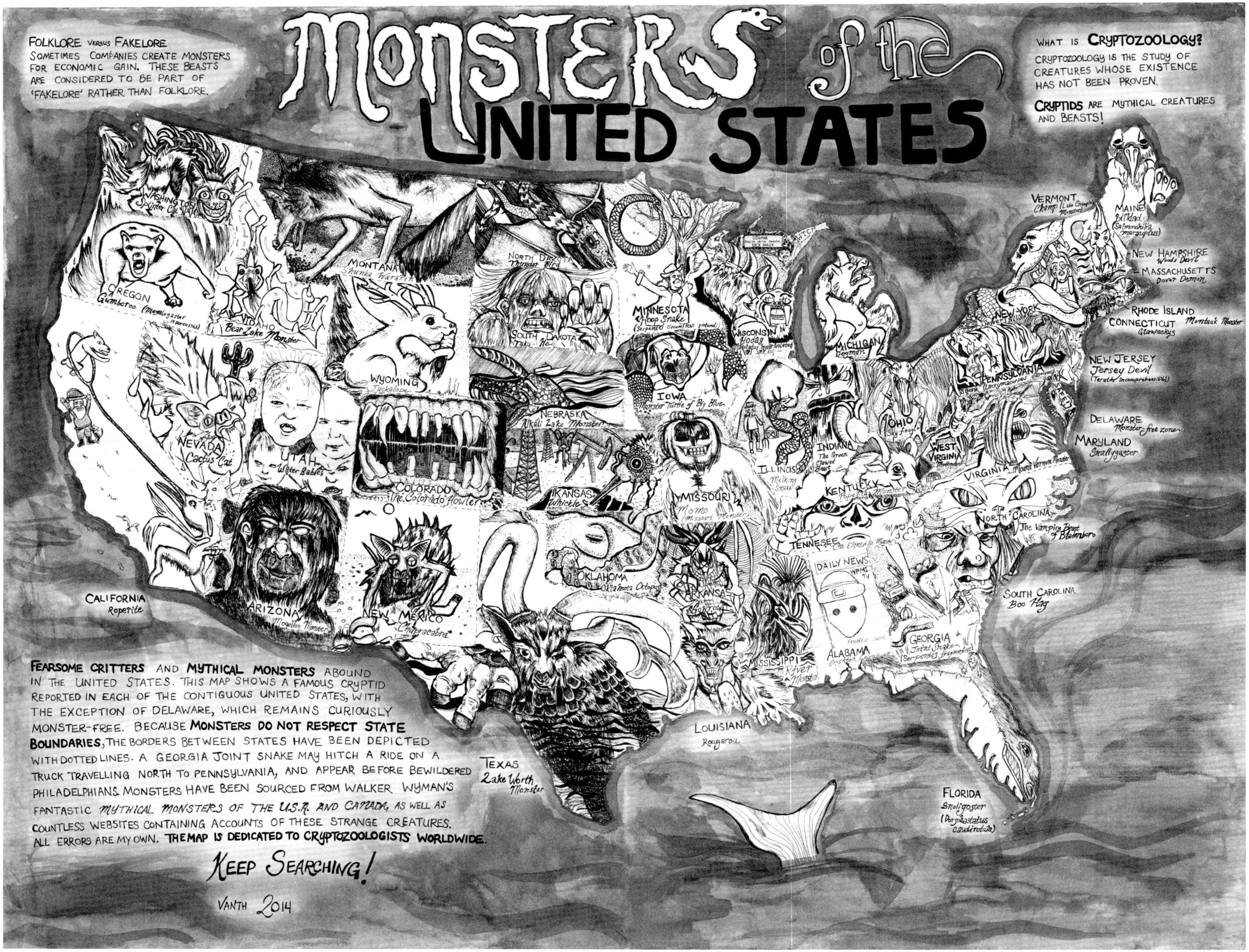

MONSTERS of the UNITED STATES
FOLKLORE versus FAKELORE
SOMETIMES COMPANIES CREATE MONSTERS FOR ECONOMIC GAIN. THESE BEASTS ARE CONSIDERED TO BE PART OF 'FAKELORE' RATHER THAN FOLKLORE.
WHAT IS CRYPTOZOOLOGY?
CRYPTOZOOLOGY IS THE STUDY OF CREATURES WHOSE EXISTENCE HAS NOT BEEN PROVEN.
CRYPTIDS ARE MYTHICAL CREATURES AND BEASTS!
WASHINGTON
OREGON
Gumberoo
IDAHO
Bear Lake Monster
MONTANA
NORTH DAKOTA
Thunder Bird
SOUTH DAKOTA
MINNESOTA
Hoop Snake
WISCONSIN
Hodag
MICHIGAN
Dogman
WYOMING
Jackalope
NEVADA
Cactus Cat
UTAH
Water Babies
COLORADO
The Colorado Howler
NEBRASKA
Alkali Lake Monster
IOWA
Monster Turtle of Big Blue
KANSAS
MISSOURI
Momo
(Missouri Monster)
ILLINOIS
Milking Snake
INDIANA
The Green Clawed Beast
OHIO
WEST VIRGINIA
Mothman
KENTUCKY
VIRGINIA
NORTH CAROLINA
The Vampire Beast of Bladenboro
TENNESEE
OKLAHOMA
Oklahoma Octopus
ARKANSAS
MISSISSIPPI
River Monster
ALABAMA
DAILY NEWS
GEORGIA
Joint Snake
SOUTH CAROLINA
Boo Hag
FLORIDA
Snoligoster
LOUISIANA
Rougarou
TEXAS
Lake Worth Monster
NEW MEXICO
Chupacabra
ARIZONA
Mogollon Monster
CALIFORNIA
Roperite
VERMONT
Champ (Lake Champlain Monster)
MAINE
NEW HAMPSHIRE
Woods Devil
MASSACHUSETTS
Dover Demon
RHODE ISLAND
Montauk Monster
CONNECTICUT
Glawackus
NEW YORK
PENNSYLVANIA
NEW JERSEY
Jersey Devil
DELAWARE
Monster-free zone
MARYLAND
Snallygaster
FEARSOME CRITTERS AND MYTHICAL MONSTERS ABOUND IN THE UNITED STATES. THIS MAP SHOWS A FAMOUS CRYPTID REPORTED IN EACH OF THE CONTIGUOUS UNITED STATES, WITH THE EXCEPTION OF DELAWARE, WHICH REMAINS CURIOUSLY MONSTER-FREE. BECAUSE MONSTERS DO NOT RESPECT STATE BOUNDARIES, THE BORDERS BETWEEN STATES HAVE BEEN DEPICTED WITH DOTTED LINES. A GEORGIA JOINT SNAKE MAY HITCH A RIDE ON A TRUCK TRAVELLING NORTH TO PENNSYLVANIA, AND APPEAR BEFORE BEWILDERED PHILADELPHIANS. MONSTERS HAVE BEEN SOURCED FROM WALKER WYMAN'S FANTASTIC MYTHICAL MONSTERS OF THE U.S.A. AND CANADA, AS WELL AS COUNTLESS WEBSITES CONTAINING ACCOUNTS OF THESE STRANGE CREATURES. ALL ERRORS ARE MY OWN. THE MAP IS DEDICATED TO CRYPTOZOOLOGISTS WORLDWIDE.
KEEP SEARCHING!
VANTH 2014

« Monsters of the United States

Chelsea Nestel

The idea for this map came to me in October of 2014, as I listened to the song "Monster Mash" by Bobby "Boris" Pickett & The Crypt-Kickers. The song is about a monster party that began spontaneously in a scientist's laboratory. As Boris crooned, I thought, "Why not have a monster party in the cartography lab at UW-Madison? I could make a Monster Map!"

First I researched which monsters to invite onto my map. I visited the Wisconsin Historical Society and checked out several books by Wisconsin folklorist Walter Wyman, who studied myths about monsters across America. With Wyman's help, I identified a representative monster for most of the 48 contiguous states. Of course, monsters do not respect political boundaries*, so I drew state borders as dotted lines to indicate that these boundaries are porous.

Wyman's research dated to the 1970s, so I scoured the Internet for more recent reports of monster activity and discovered a number of fresh monsters that had become part of American folklore. The Crichton Leprechaun, for example, became famous through a news item in 2008. In addition to gathering their stories, I researched visual depictions of the monsters because I knew the cryptids (creatures whose existence has been suggested but not scientifically verified) would want to look their best.

Armed with addresses and mug shots of the monsters, I began drawing. I used the traditional cartooning techniques of permanent ink and wash on watercolor paper. The original map took three weeks to produce, with dimensions of 48 x 36 inches. To digitize the map, I scanned the original and made small adjustments in Photoshop to enhance the clarity of the text. I kept digital alterations to a minimum to preserve the hand-produced quality of the map.

I hope that this map will give viewers a reason to talk to each other and compare cryptids: "Hey! You're from California. What's the deal with the Roperite?" I also hope they will become interested in investigating the strange, scary, and fascinating folklore of the United States. Let's all do the Monster Mash!

** Delaware is the exception to this rule. In addition to being a sales-tax-free state, Delaware is a monster-free zone! Despite a few weeks of research, I could not find a single monster that originated in Delaware.*

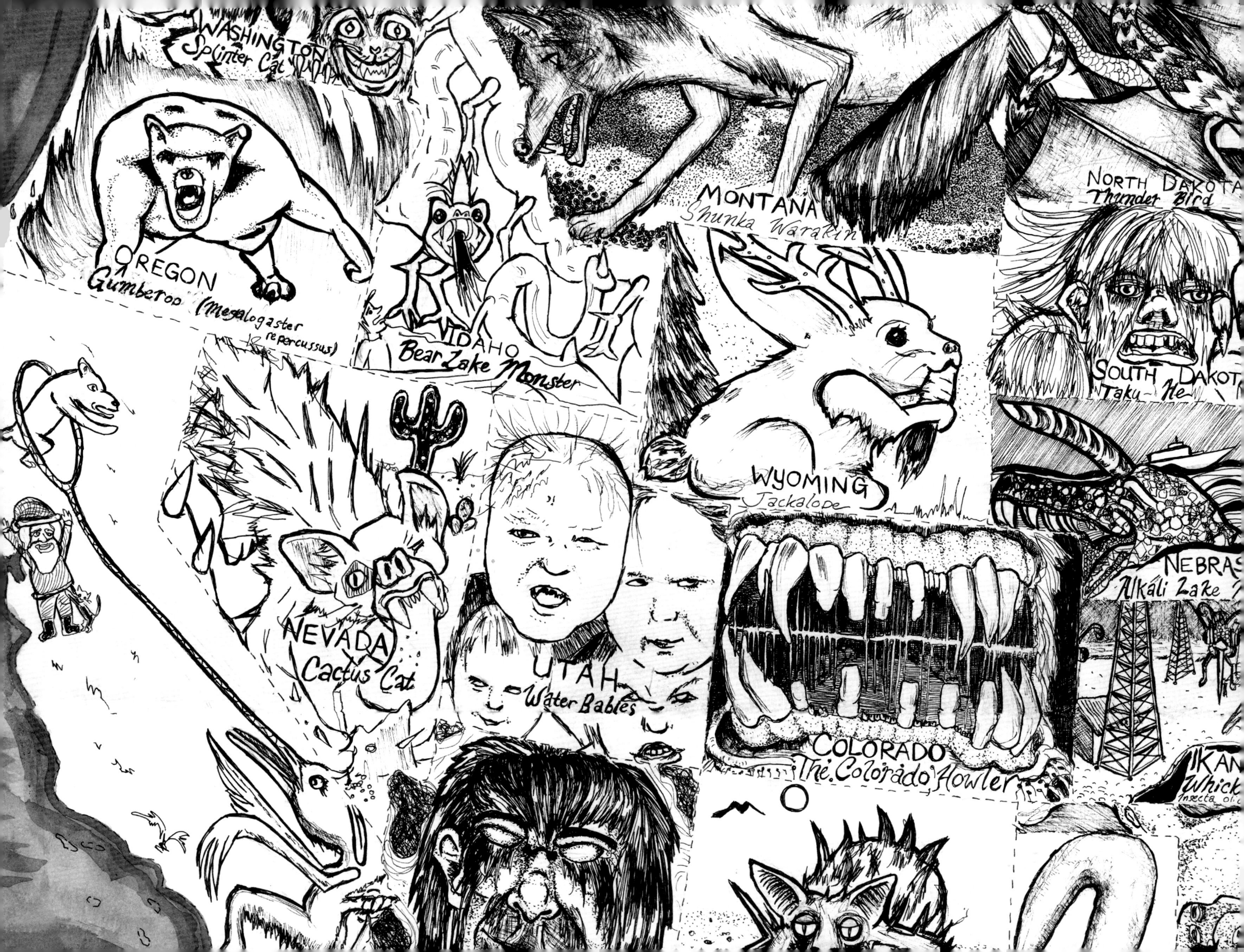
WASHINGTON
Splinter Cat
OREGON
Gumberoo (Megalogaster repercussus)
IDAHO
Bear Lake Monster
MONTANA
Shunka Warakin
NORTH DAKOTA
Thunder Bird
SOUTH DAKOT
Taku He
WYOMING
Jackalope
NEVADA
Cactus Cat
UTAH
Water Babies
COLORADO
The Colorado Howler
NEBRAS
Alkali Lake
KAN
Whick

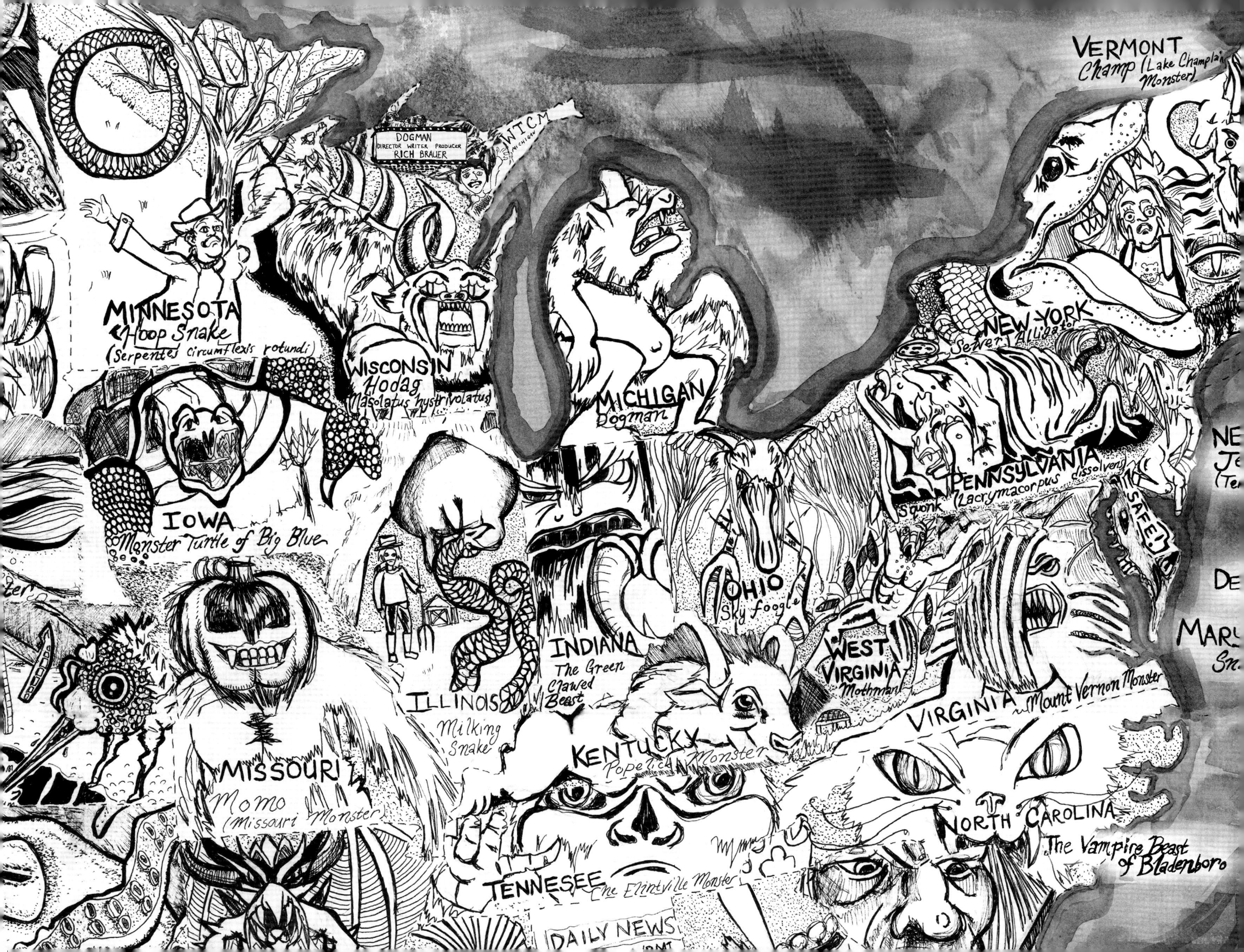
VERMONT
Champ (Lake Champlain Monster)
DOGMAN
DIRECTOR WRITER PRODUCER
RICH BRAUER
MINNESOTA
Hoop Snake
(Serpentes circumflexis rotundi)
WISCONSIN
Hodag
(Masolatus hystrivolatus)
MICHIGAN
Dogman
NEW YORK
Sewer Alligator
PENNSYLVANIA
Squonk (Lacrymacorpus dissolvens)
SAFE!
IOWA
Monster Turtle of Big Blue
OHIO
Sky foogle
INDIANA
The Green Clawed Beast
WEST VIRGINIA
Mothman
VIRGINIA
Mount Vernon Monster
ILLINOIS
Milking Snake
KENTUCKY
Pope Lick Monster
MISSOURI
Momo
(Missouri Monster)
NORTH CAROLINA
The Vampire Beast of Bladenboro
TENNESEE
The Flintville Monster
DAILY NEWS